Principles of
Genetic Toxicology

Principles of Genetic Toxicology

David Brusick
Litton Bionetics
Kensington, Maryland

PLENUM PRESS · NEW YORK AND LONDON

Library of Congress Cataloging in Publication Data

Brusick, David.
  Principles of genetic toxicology.

  Includes bibliographies and index.
  1. Chemical mutagenesis. 2. Mutagenicity testing. 3. Teratogenic agents. I. Title.
[DNLM: 1. Environmental pollutants. 2. Mutagens. 3. Mutation. QH465.C5 B912p]
QH465.C5B78   573.2'292   80-16514
ISBN 0-306-40414-1

© 1980 Plenum Press, New York
A Division of Plenum Publishing Corporation
227 West 17th Street, New York, N.Y. 10011

To

Herman Brockman

Geneticist, Scholar, and Teacher.
Without his influence and guidance,
this book would not have been written.

Foreword

The field of genetic toxicology is a relatively new one which grew out of the studies of chemical mutagenesis and modern toxicology. Considering that systematic practices to detect chemical mutagenesis are only a little over thirty years old, this field has evolved very rapidly with an abundance of methods for identifying chemical mutagens.

To evalulate the usefulness of the methods and to select the assay which will yield the most important information under practical conditions requires the broad experience such as that which Dr. Brusick has acquired over the last decade. Since this field is expanding very rapidly and new testing methods are being recognized, it should be kept in mind that revisions may have to be made during the next five to ten years.

The need for such a book has been obvious to us, particularly since training courses and workshops on genetic toxicology are being organized which find it beneficial to utilize established guidelines and since the reports in the literature do not always describe in detail how the work was carried out during laboratory testing.

In addition to his broad background in genetics, Dr. Brusick has had much practical experience, having organized and directed the most extensive laboratory for mutagen testing. I am most pleased to see this volume, for which there is an ever-increasing need.

Alexander Hollaender
Associated Universities, Inc.
1717 Massachusetts Avenue, N.W.
Washington, D.C. 20036

Preface

The present generation is only a caretaker of the human genome of future generations.

—Malling and Valcovic (1978)*

DNA is well over a billion years old and was the basic informational macromolecule for the evolution of all life forms in existence today. The antiquity of DNA binds all living organisms to one another, for we as humans have many genes in common with plants, animals, and microbes. Because development of common genes must predate the point at which the evolutionary paths for plants and animals separated, many of these genes are hundreds of millions of years old.

During the course of the evolution of the human organism, the configuration and the number of genes per cell changed due to mutation and various types of recombinational events. The composition of the total gene pool determines the genetic integrity of our species. Our generation has temporary custody of the gene pool, and all future humans will be derived from this repository. The gene pool was not received unblemished, for there are numerous altered genes that produce hereditary diseases. It is, however, our responsibility to prevent further increases in that mutational burden.

Great strides in industrialization, chemical synthesis, and nuclear energy have to a large extent occurred during the past 70 years. With these strides have come benefits as well as the inevitable risks. Some of these risks are in the form of man-made mutagens, and we are only beginning to understand the possible impact of these mutagens on the human gene pool. Mutagens, both naturally occurring and man-made, are present in our envi-

* See reference 21 in Chapter 1.

ronment. The importance of these agents on human health could be profound if the extent of exposure is sufficient to produce genotoxic effects in critical cellular targets.

Genetic toxicology is the discipline that addresses this problem. It is a relatively new entity among the mixture of disciplines that make up the science of toxicology, but it draws upon an experimental data base in genetic studies spanning over 100 years. The continued advancement of genetic toxicology is important to the future well-being of the human species. The increased mutational load that might be added to the gene pool through our lack of concern or our insensitivity to the problems of mutation will penalize all succeeding generations. It should be our goal to prevent the present generation's being identified as an unconcerned caretaker.

The purpose of this book is to provide a synthesis of genetics and toxicology. Many geneticists who wish to move into an applied science need to understand the orientation of toxicology and its goals; and likewise, toxicologists or scientists trained in other areas of industrial health and safety need to become acquainted with the fundamentals of genetics as a basic and applied science. This is a relatively new venture, but as applied health sciences begin to overlap more areas of basic science, this type of synthesis will become more common.

David J. Brusick

Kensington, Md.

Contents

Chapter 9
SAMPLE STUDY DESIGNS 191

Origins of Genetic Toxicology

INTRODUCTION

The field of toxicology deals with effects of agents on living systems with the overt or in some cases implied purpose of defining human health effects. It is an applied science which draws on data and methodology from a multitude of basic sciences such as physiology, pharmacology, metabolism, ethology, genetics, embryology, chemistry, and statistics.

Over the past 20 years, toxicology has provided the primary source of data on health effects for chemical safety evaluations on existing and new products. Without this information, many potentially hazardous chemicals would have been identified only through human use and experience. While human experience via accidental exposure, war, and approved human experimentation (e.g., experimental chemotherapy on the terminally ill) has provided some of the most relevant data to toxicologists, this approach to routine testing of new chemicals prior to their introduction into the environment is unacceptable.

Alternatively, animal models are used in toxicologic evaluations. Mammals are the most common test organisms employed. The choice of a mammalian species depends on two factors: its suitability as a model for the human experience, that is, its close resemblence to humans in its response to selected agents; and economic considerations such as availability and cost of the animal. For the assessment of acute toxicity and for reproduction studies, rodent species appear suitable, but for specific types of toxicity only certain mammals respond in a manner similar to humans. This situation is particularly true for psychopharmacology and studies of toxicity associated with nutrition or aging.[10,17]

Numerous subspecialties of toxicology must rely on inappropriate model systems because suitable human surrogates are not available. Examples of areas lacking good model systems are teratology, skin sensitization, dermal adsorption and toxicity, and possibly certain types of carcinogenesis.[8,27,34] Thus, toxicology is an inexact field encumbered with uncertain extrapolation of effects observed in animal models exposed to chemicals under laboratory conditions that do not always duplicate environmental exposures.

Reliable estimates of human risk depend upon an understanding of the cellular and molecular mechanisms that underlie toxic end points expressed at the whole organism level, and on adequate methods of quantitating both the real dose of a chemical (amount reaching the critical target site) and the level of the response. Dose is most often equated with exposure, that is, the environmental concentration surrounding the test organism, which seldom represents the dose delivered to the critical target.

Animal toxicology can be divided into many subspecialties, and each subspecialty can be a separate field of study. The unifying superstructure among various subdivisions of toxicology appears to be the duration of the study. Most studies fall under one of three categories: acute, subchronic, or chronic. The duration of each type of study depends on the test organism, but in general for small rodents acute studies run from 1 to 15 days, subchronic studies from 15 to 120 days, and chronic studies from 12 months to the lifetime to the animal. Table 1.1 summarizes some of the typical toxic end points measured and the duration of the study associated with them.

THE HISTORY OF GENETIC TOXICOLOGY

Genetic toxicology as a subspecialty of toxicology identifies and analyzes the action of agents with toxicity directed towards the hereditary components of living systems. While many toxicants damage hereditary material once they reach a level which produces generalized nonspecific cell toxicity and death, it is the primary objective of genetic toxicology to detect and understand the properties of the relatively small group of agents which are highly specific for nucleic acids and produce deleterious effects in genetic elements at subtoxic concentrations. Compound exposures for genetic toxicology studies range from acute to chronic, and as such this type of testing falls into the three major temporal subdivisions of toxicologic testing.

Agents specifically producing genetic alterations at subtoxic exposure levels which result in organisms with altered hereditary characteristics are called genotoxic. Genotoxic substances usually have chemical or physical

TABLE 1.1

Typical Toxic End Points or Signs Observed in the Major Temporal Subdivisions of
Animal Toxicology

Acute	Subchronic[a]	Chronic[b]
Lethality	Reproductive capacity	Nonreversible tissue
Irritation		degeneration
Necrosis	Teratology	Carcinogenesis
Changes in normal homeostasic	Reversible tissue	Life expectancy
parameters	degeneration	
Neurological effects	Dietary deficiencies	
	Behavioral changes	

[a] Subchronic effects may also include all of those listed under Acute.
[b] Chronic effects may also include those listed under Acute and Subchronic.

properties that facilitate interaction with nucleic acids. In fact, it is the universality of the target molecule that provides the scientific basis for the discipline of genetic toxicology. A derivation of the term genotoxic is found in reference 15.

The beginnings of genetic toxicology could probably be said to have evolved from the initial studies of gene mutability demonstrated first by Muller in 1927 using radiation,[23] followed almost 20 years later by Auerbach using chemicals.[3] Both of these investigators conducted their studies using submammalian species, but within the next 20 years, genetic changes in animals induced by radiation and chemicals were demonstrated by Cattanach and by Russell at Oak Ridge.[11,25] This work created the awareness that some of the "hereditary" diseases observed in human populations might be environmental in origin.[6,26] Proof, in the early 1940s, that deoxyribonucleic acid (DNA) is the hereditary material, and the subsequent elucidation of its primary, secondary, and tertiary structures by Watson and Crick in the early 1950s, opened up new avenues of research into the mechanisms of mutagenesis.

The period of time from 1953 to 1968 was the "Golden Era" of molecular genetics. During this time much of the basic information was developed regarding DNA structure and replication, the genetic code, mechanisms of protein synthesis, and DNA repair processes (Table 1.2). Cell biologists and biochemists and microbiologists reigned over this Golden Era, and several received Nobel Prizes for their contributions.

Genetic toxicology was recognized as a discipline around 1969. The Environmental Mutagen Society was founded in that year under the leadership of Dr. Alexander Hollaender and several other geneticists who were concerned about the potential genetic impact associated with the pro-

TABLE 1.2
Major Discoveries in Molecular Genetics

Phenomenon	Year	Reference
DNA constitutes the hereditary material	1944	4
Elucidation of DNA structure	1953	32
Mechanism of protein synthesis— the central dogma hypothesis	1957	12
Demonstration of semiconservative replication of the DNA molecule	1958	22
Operon model of gene regulation	1961	18
Elucidation of the genetic code for protein	1964–1967	24
Enzymatic synthesis of DNA *in vitro*	1967	13

liferation of man-made chemicals in the environment. The concept of genetic toxicology was clearly consistent with the intense concern for environmental protection that prevailed at the time.

Occurring almost simultaneously with the growing concern by geneticists over environmental mutagens were the reports by several groups of investigators showing a correlative relationship between mammalian carcinogens and mutagens.[1,5,9,21,28] Several earlier attempts to correlate the properties of mutagenesis and carcinogenesis failed because of the limitations inherent in the genetic assays available.[7] However, following the introduction of carcinogen activation using host-mediated[16] and *in vitro* microsome activation systems[19] developed from 1969 through 1971, the concept of mutagenic carcinogens was revitalized.

Genetic toxicology, therefore, plays a dual role in overall health effects studies. One function is the implementation of testing and risk assessment methods to define the impact of genotoxic agents that are found in the environment and whose presence may alter the integrity of the human gene pool. The second function is the elucidation of the relationship between genotoxicity and the initiation of neoplasia. In this latter regard, genetic toxicology has been applied as a front-line screen for chemicals with properties which might cause them to be carcinogens.

During the past 5 years, genetic toxicology has continued to grow as a legitimate specialty area of toxicology, and the influence of this discipline on overall chemical safety evaluation has been significant. Specific guidelines or regulations detailing testing for potential genotoxic agents are in effect or under consideration by all regulatory agencies of the United States Government and several other nations.

TECHNOLOGY TRANSFER TO APPLIED GENETICS

The application of fundamental genetic methodology to human problems occurred prior to the recent concern over genotoxicity. Plant and animal breeding are both derived from early experimentation in genetics. This applied science is an international multibillion-dollar business involving the development of new varieties of plants and animals for both recreational and agricultural purposes.

Technology transfer, the application of basic science methodology to the solution of practical problems, has been essential to the development of genetic toxicology. Many of the methods employed to detect genotoxic substances were initially developed for other purposes. The Ames *Salmonella*/microsome assay, for example, evolved from early biochemical investigations of histidine biosynthesis in *Salmonella typhimurium*.[33] This test method, as well as others—such as those used to measure stimulation of DNA repair processes (unscheduled DNA synthesis)—was originally part of research efforts directed toward the elucidation of DNA repair pathways. Technology transfer is still proceeding rapidly in genetic as well as other areas of toxicology; in fact, a minor revolution is occurring.

Toxicology has traditionally been an animal testing science relying primarily on the detection of qualitative and quantitative changes in animal behavior, homeostatic processes, and lethality. From such observations, a toxicity profile of a substance is generated. This profile is then used to assess the probable consequences following human exposure to the substance, and it often forms the basis for regulating environmental levels of a material. Thus, toxicology can be characterized as a descriptive science since often little is known regarding the molecular events underlying the expression of the toxicologic end point measured.

As a consequence of numerous scientific and regulatory pressures, interest in elucidating the molecular mechanisms for all forms of toxicity has increased. This new approach to toxicologic investigation has been termed molecular toxicology. Molecular toxicologists seek to define mechanisms and underlying fundamental processes. Genetic toxicology, behavioral toxicology, immunotoxicology, teratology, and oncogenesis are subspecialties of toxicology in which scientists are attempting to relate these specific end points to molecular mechanisms.

One important factor stimulating the development of molecular toxicology is the increasing costs associated with animal testing. These costs continue to rise at the same time the necessity for evaluating the toxicity of more chemicals is increasing. Thus, searches for rapid, economical tests which identify substances as probable toxicants based on their interaction with key molecular processes should continue to expand over the next several years.

THE COMPONENTS OF GENETIC TOXICOLOGY

The results of technology transfer have produced a remarkable turnover in genetic toxicology assay systems during the past several years. New test systems are evaluated for their suitability, and those with promise are often incorporated into the array of tests employed in this science. Unfortunately, as new test systems have been added, few have been retired. Thus, genetic toxicology is cluttered with tests of limited value and relevance for real genotoxic assessment. For example, tests for phage induction (induct test) and biphenyl hydroxylation appeared promising in early phases of validation, but they have exhibited low levels of compound discrimination when tested with a wider range of chemical classes and are not generally considered reliable indicators of genotoxicity. It is possible that these tests will have value in restricted situations. Another test, the host-mediated assay, originally described by Gabridge and Legator,[16] is still used, but has undergone modifications which have significantly enhanced its utility compared with the original methodology.[2,14] The original purpose of host-mediated techniques was to provide a means for metabolic activation of promutagen agents. However, the demonstration by Malling in 1971 that *in vitro* microsome metabolic systems could be coupled with microbial mutagenicity testing essentially replaced host-mediate methods as screening tests.[19]

Some types of test methods which had lost popularity with the introduction of S9-supplemented microbial and mammalian cell culture tests are being rediscovered. Extensive research efforts involving *Drosophila melanogaster* have renewed interest in this classic genetic test organism. *Drosophila* has many advantages, including its own metabolic activation system and the ability to screen for both gene and chromosome effects.[31]

New tests currently under examination using mammals offer the possibility of measuring a broad array of genetic end points *in vivo*. These tests encompass both somatic and germ cell target sites.[20]

Thus, the entire array of tests used to detect genotoxic compounds covers a broad range of techniques, including some relatively unusual types of assays such as the silk worm oocyte assay[29] and mutant induction in *Tradescantia* stamen hairs[30] as well as more common assays such as the dominant lethal test derived from traditional toxicology techniques. However, only a relatively small number of the total array of tests are presently included in routine screening programs found in toxicology testing laboratories.

The route of new tests from the experimental to the applied stages generally follows a similar pattern (Figure 1.1). The first step is development and confirmation of a technique which is often derived from funda-

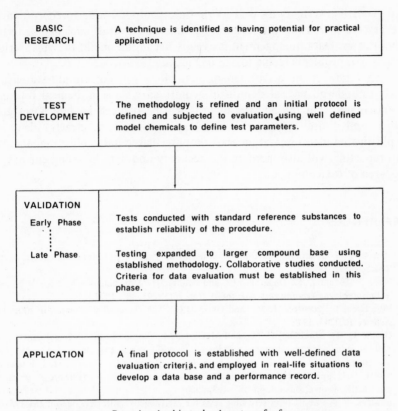

FIGURE 1.1. Steps involved in technology transfer for new assays.

mental research. It is at this step that the verification of the nature of the end point is determined (i.e., the end point must be a true genetic alteration, not a phenomenon resembling a genetic event). The basic methodology and protocol are also established during the development and verification stage. The development process usually employs only model chemicals for which a great deal of information regarding their mechanisms of action is already known. The second step is one of validation. The overall utility of an assay rests on its ability to detect genetic activity in a wide range of chemical classes. Once a relatively standardized protocol is established, an assay is subjected to vigorous evaluation in multiple laboratories using coded samples. The final step is the application of the assay in "real life" testing with a uniform protocol.

As in other areas of toxicology, a secondary type of technology transfer is gaining momentum in genetic toxicology. Computer-based data handling and reporting systems as well as on-line, computer-monitored equipment are greatly increasing the turnaround time and overall capacity of testing laboratories. Data management is rapidly becoming one of the most serious generalized problems facing toxicology testing.

The state of the art in genetic toxicology will not remain static; it is likely to undergo several significant evolutionary steps in the near future. It is anticipated that the number of tests will be reduced as data bases grow larger and those tests with serious limitations are clearly identified. Sophisticated instrumentation and computer-based data collection, analysis, and reporting will also increase the accuracy and data handling capacity of this area of toxicology.

REFERENCES

1. Ames, B. N.: Identifying environmental chemicals causing mutations and cancer. *Science* **204**:587–593, 1979.
2. Arni, P., Mantel, T., Deparade, E., and Muller, D.: Intrasanguine host-mediated assay with *Salmonella typhimurium*. *Mutation Res.* **45**(3):291–307, 1977.
3. Auerbach, C., Robson, J. M., and Carr, J. G.: The chemical production of mutations. *Science* **105**:243, 1947.
4. Avery, O. T., MacLeod, C. M., and McCarty, M.: Studies on the chemical nature of the substance inducing transformation of Pneumococcal types. *J. Exp. Med.* **79**:137–158, 1944. Reprinted in *Classic Papers in Genetics* (J. A. Peters, ed.), Prentice-Hall, Englewood Cliffs, N.J., pp. 173–192, 1959; and *Papers on Bacterial Genetics* (Edward A. Adelberg, ed.), Little, Brown, Boston, pp. 147–168, 1960.
5. Bridges, B. A.: Short-term screening tests for carcinogens. *Nature (London)* **261**:195–200, 1976.
6. Bridges, B. A., Bochkov, N. P., and Jansen, J. D.: Genetic monitoring of human populations accidentally exposed to a suspected mutagenic chemical. International Commission for Protection against Environmental Mutagens and Carcinogens, Publication No. 1, *Mutat. Res.* **64**:57–60, 1979.
7. Brusick, D. J.: *In vitro* mutagenesis assays as predictors of chemical carcinogenesis in mammals. In *Toxicology Annual*, Vol. 2 (Charles L. Winek, ed.), Marcel Dekker, New York, pp. 79–105, 1977.
8. Brusick, D. J.: The value and significance of carcinogenic, mutagenic, and teratogenic tests. In *Cutaneous Toxicity* (V. A. Drill and P. Lazar, eds.), Academic Press, New York, pp. 189–201, 1977.
9. Brusick, D. J.: The role of short-term testing in carcinogen detection. *Chemosphere* No. 5, pp. 403–417, 1978.
10. Burnet, F. M.: *Intrinsic Mutagenesis: A Genetic Approach to Aging*, Medical and Technical Publishing, Lancaster, England, 1974.
11. Cattanach, B. M.: Chemically induced mutations in mice. *Mutat. Res.* **3**:346–353, 1966.
12. Crick, F. H. C.: On protein synthesis. *Symp. Soc. Exp. Biol.* **12**:138–163, 1958.
13. Goulian, M., Kornberg, K., and Sinsheimer, R. L.: Enzymatic synthesis of DNA, XXIV. Synthesis of infectious phage ΟΧ174 DNA. In *Molecular Biology of DNA and RNA: An*

Analysis of Research Papers (I. D. Raacke. ed.), C. V. Mosby, St. Louis, pp. 112–119, 1971.

14. Durston, W. E., and Ames, B. N.: A simple method for the detection of mutagens in urine: Studies with the carcinogen 2-acetylaminofluorene. *Proc. Natl. Acad. Sci. U.S.A.* **71**:737–741, 1974.

15. Ehrenberg, L., Brookes, P., Druckrey, H., Lagerlof, B., Litwin, J., and Williams, G.: The relation of cancer induction and genetic damage. In *Evaluation of Genetic Risks of Environmental Chemicals*, Report of Group 3, Ambio Special Report No. 3, Royal Swedish Academy of Sciences, Universitetsforlaget, 1973.

16. Gabridge, M. G., and Legator, M. S.: A host-mediated assay for the detection of mutagenic compounds. *Proc. Soc. Exp. Biol. Med.*, **130**:831, 1969.

17. Hanin, I., and Usdin, E. (eds.): *Animal Models in Psychiatry and Neurology*, Pergamon Press, New York, 1978.

18. Jacob, F., and Monod, J.: On the regulation of gene activity. *Cold Spring Harbor Symp. Quant. Biol.* **26**:193, 1961.

19. Malling, H. V.: Dimethylnitrosamine: Formation of mutagenic compounds by interaction with mouse liver microsomes. *Mutat. Res.* **13**:425, 1971.

20. Malling, H. V., and Valcovic, L. R.: New approaches to detecting gene mutations in mammals. In *Advances in Modern Toxicology*, Vol. 4 (G. Flamm and M. Mehlman, Eds.), Halsted Press, New York, Chapter 8, 1978.

21. McCann, J., Choi, E., Yamasaki, E., and Ames, B. N.: Detection of carcinogens as mutagens in the *Salmonella*/microsome test: Assay of 300 chemicals. *Proc. Natl. Acad. Sci. U.S.A.* **72**(12):5135–5139, 1975.

22. Meselson, M. S., and Stahl, F. W.: The replication of DNA in *Escherichia coli. Proc. Natl. Acad. Sci. U.S.A.* **44**:671–682, 1958.

23. Muller, H. J.: Artificial transmutation of the gene. *Science* **66**:84–87, 1927.

24. Nirenberg, M., Caskey, T., Marshall, R., Brimacombe, R., Kellogg, D., Doctor, B., Hatfield, D., Levin, J., Rottman, F., Pestka, S., Wilcox, M., and Anderson, F.: The RNA code and protein synthesis. In *Papers on Genetics: A Book of Readings* (Louis Levine, ed.), C. V. Mosby, St. Louis, pp. 34–52, 1971.

25. Russell, W. L.: X-ray induced mutations in mice. *Cold Spring Harbor Symp. Quant. Biol.* **16**:327–330, 1951.

26. Russell, W. L.: The role of mammals in the future of chemical mutagenesis research. *Arch. Toxicol.* **38**:141–147, 1977.

27. Shubik, P.: Identification of environmental carcinogens: Animal test models. In *Carcinogens: Identification of Mechanisms of Action* (A. C. Griffin and C. R. Shaw, eds.), Raven Press, New York, pp. 37–47, 1979.

28. Sugimura, T., Sato, S., Nagao, M., Yahagi, T., Matsushima, T., Seino, Y., Takeuchi, M., and Kawachi, T.: Overlapping of carcinogens and mutagens. In *Fundamentals in Cancer Prevention* (P. N. Magee, T. Matsushima, T. Sugimura, and S. Takayama, eds.), University of Tokyo Press, Tokyo, and University Park Press, Baltimore, Maryland, pp. 191–215, 1976.

29. Tazima, Y., and Onimaku, K.: Results of mutagenicity testing for some nitrofuran derivatives in a sensitive test system with silkworm oocytes. Japanese Environmental Mutagen Society, 2nd Annual Meeting, Abstract 16.

30. Underbrink, A. G., Schairer, L. A., and Sparrow, A. H.: *Trandescantia* stamen hairs: A radiobiological test system applicable to chemical mutagenesis. In *Chemical Mutagens: Principles and Methods for Their Detection*, Vol. 3 (A. Hollaender, ed.), Plenum Press, New York, Chapter 30, p. 171, 1976.

31. Vogel, E., and Sobels, F. H.: The function of *Drosophila* in genetic toxicology testing. In *Chemical Mutagens: Principles and Methods for Their Detection*, Vol. 4 (A. Hollaender, ed.), Plenum Press, New York, Chapter 38, pp. 93–142, 1976.
32. Watson, J. D., and Crick, F. H. C.: The structure of DNA. In *Papers on Genetics: A Book of Readings* (Louis Levine, ed.), C. V. Mosby, St. Louis, pp. 11–21, 1971.
33. Whitfield, H. J., Martin, R. G., and Ames, B. N.: Classification of aminotransferase (C gene) mutants in the histidine operon. *J. Mol. Biol.* **21**:335, 1966.
34. Wilson, J. G.: *Environment and Birth Defects*, Academic Press, New York, Chapter 8, 1973.

CHAPTER 2

Fundamentals of Genetic Toxicology

INTRODUCTION

The purpose of this chapter is to describe the genetic background and terminology essential for an understanding of genetic toxicology. It is necessary to appreciate the basic structure of the target molecule, DNA, and to understand how it operates, since the types of molecular lesions that chemicals induce in this molecule and the genetic effects that they generate are intimately tied to the structure and function of this molecule. Because of the unique position held by DNA in maintaining and processing cellular information, intrinsic capacity for self-repair appears to have evolved simultaneously with the environmental adaptation of this molecule. Repair processes are also believed to influence the kinetics of mutation induction by preventing many chemically induced lesions from becoming fixed as permanent alterations.

The first portion of this chapter presents some of the fundamentals of the science of genetics. The second section is devoted to a presentation of the categories and mechanisms of DNA alterations which generate genotoxic effects in cells.

BASIC GENETICS FOR TOXICOLOGISTS

Gene Structure

DNA is the macromolecule from which all characteristics of life emanate (Figure 2.1). The informational molecules of all living systems,

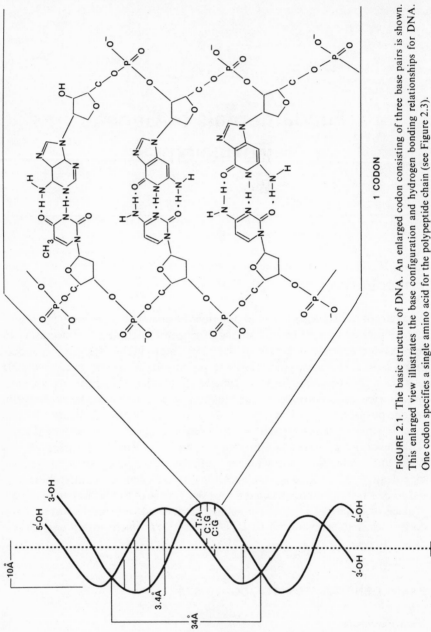

1 CODON

FIGURE 2.1. The basic structure of DNA. An enlarged codon consisting of three base pairs is shown. This enlarged view illustrates the base configuration and hydrogen bonding relationships for DNA. One codon specifies a single amino acid for the polypeptide chain (see Figure 2.3).

plant cells.[8] Universality in all phases of gene decoding is fundamental to success of this type of genetic research.

Gene Function

With only minor exceptions, it is assumed that gene function at the molecular level parallels DNA structural similarity and is identical in all organisms. The following characteristics represent the basics of gene function.

1. A gene is composed of a segment of a DNA molecule containing sufficient information to produce a functional product. The products of almost all genes are polypeptides. The function of these polypeptides can be structural, enzymatic, or regulatory. All visible and functional qualities of an organism, defined as the phenotype, are dependent upon the accurate structure and expression of these gene products.

2. The information required to specify a single amino acid is contained in a triplet of base pairs called a codon. The sequence of the bases in the codons is important for translation of information. Table 2.2 gives the codon dictionary for the naturally occurring amino acids used to synthesize protein. More than one codon can specify a given amino acid. For example, 5' CGU 3', 5' CGA 3', and 5' AAG 3' all code for the amino acid arginine. The sequence of codons in a gene specifies the sequence of amino acids in a polypeptide (Figure 2.2) and thus its ultimate role in cellular processes.

3. Changes in the composition of the base pair of a codon or changes in the codon composition of a gene can result in a gene product that will not function. This may eventually be expressed in cell or organism death, or in an altered form of the cell or organism. Thus, these changes form the basis of genetic alterations, or "mutations."

4. The production of a polypeptide gene product involves two types of RNA: messenger RNA (m-RNA), a short-lived copy of the gene being expressed; and transfer RNA (t-RNA), which contributes to both the translocation and coupling of amino acids into the polypeptide gene product. This latter process also involves an RNA/protein structure called a ribosome. The basic process is illustrated in Figure 2.3.

5. The initiation and termination of the transcription of a gene into a polypeptide is regulated by a separate set of regulatory genes. Most regulatory genes respond to chemical or temporal cues such that only those genes needed at a given time are expressed. The remaining genes are in an inactive state. The processes of gene activation and inactivation are believed to be critical to cellular differentiation in multicellular organisms.

Gene function and nucleic acid organization at the cellular level differ in prokaryotic and eukaryotic cell types (Figure 2.4). For example, DNA location and packaging are distinctly different in bacteria (prokaryotic) and

with the exception of some viruses which use ribonucleic a
composed of DNA. Some of the characteristic features of
are listed in Table 2.1. The fact that the DNA of all living
bacteria to humans, have these characteristics emphasizes
its origin and the immortality of those genes which pertai
housekeeping processes of DNA synthesis, replication, and
mechanisms of information storage and gene expression
identical in all organisms.

The simplest complete functional unit in a DNA mol
Most of what is known about the structure and operation
acquired from studies with bacteria or bacteriophages.[4,17,2]
between the genes of prokaryotic (bacteria) and eukaryotic
cells) organisms are minor and primarily center on their nu
on the chromosomal entity and mechanisms of gene
nucleotide composition and the mechanisms by which infor
a gene are transformed into gene products appear to b
organisms. This has been proven by the recent developmer
DNA genetic engineering, where genes continue to funct
having been transplanted from human cells to bacteria cell

TABLE 2.1

Basic Biochemical Characteristics of All Double-Strand

1. DNA consists of two different purines (guanine, adenine) and two different pyrimidines (thymine and cytosine).

2. A nucleotide pair consists of one purine and one pyrimidine [adenine/thymine (AT) or guanine/cytosine (GC)].

3. Nucleotide pairs are connected into a double helix molecule by sugar–phosphate backbone linkages and hydrogen bonding (Figure 2.1).

4. The AT base pair is held by two hydrogen bonds, while the GC is held by three.

5. The distance between each base pair in a molecule is 3.4 Å, producing 10 nucleotide pairs per turn of the DNA helix.

6. The number of adenine molecules must equal the number of thymine molecules in a DNA molecule. The same relationship exists for guanine and cytosine molecules. However, the ratio of AT to GC base pairs may vary in DNA from species to species.

7. The two strands complementary respect to the sugar–phosphat being 3'–5' and respect to the t ribose sugar.

8. DNA replicates method in separate and e for the synth mentary strand

9. The rate of DN tion during re 600 nucleotid must unwind of 3600 rp replication ra

10. The DNA conte 10^9 daltons f 10^{11} daltons f

TABLE 2.2
Dictionary of the Genetic Code for Proteins

Codon	Amino acid	Codon	Amino acid	Codon	Amino acid	Codon	Amino acid
UUU	PHE	UCU	SER	UGU	CYS	UAU	TYR
UUC	PHE	UCC	SER	UGC	CYS	UAC	TYR
UUA	LEU	UCA	SER	UGA	NONE	UAA	OCHRE
UUG	LEU	UCG	SER	UGG	TRYP	UAG	AMBER
CUU	LEU	CCU	PRO	CGU	ARG	CAU	HIS
CUC	LEU	CCC	PRO	CGC	ARG	CAC	HIS
CUA	LEU	CCA	PRO	CGA	ARG	CAA	GLU-NH$_2$
CUG	LEU	CCG	PRO	CGG	ARG	CAG	GLU-NH$_2$
AUU	ILEU	ACU	THR	AGU	SER	AAU	ASP-NH$_2$
AUC	ILEU	ACC	THR	AGC	SER	AAC	ASP-NH$_2$
AUA	ILEU	ACA	THR	AGA	ARG	AAA	LYS
AUG	MET	ACG	THR	AGG	ARG	AAG	LYS
GUU	VAL	GCU	ALA	CGU	GLY	GAU	ASP
GUC	VAL	GCC	ALA	GGC	GLY	GAC	ASP
GUA	VAL	GCA	ALA	GGA	GLY	GAA	GLU
GUG	VAL	GCG	ALA	GGG	GLY	GAG	GLU

[a] U substitutes for T in RNA.
[b] The DNA code sequence would be complementary to each RNA code [e.g., (DNA sequence/CAG) → (RNA sequence/GUC) → (amino acid/valine)].
[c] Codon words read from 5'-OH end (left) to 3'-OH end (right).
[d] Codons UGA, UAA, and UAG are all terminating sequences and do not code for any amino acids.
[e] All amino acids are identified by the first three letters of their name with the exception of isoleucine = ILEU, asparagine = ASP-NH$_2$, and glutamine = GLU-NH$_2$.

mammalian (eukaryotic) somatic cells. These differences will more likely influence gene expression mechanisms than mutation induction mechanisms, and as such should not greatly interfere with extrapolation of genotoxic effects across phylogenetic barriers. The following series of statements represents the basics of nucleic acid organization at the cellular level.

1. DNA may be packaged with proteins into chromatin material and organized into chromosomes, or it may be present as an uncomplexed DNA helix as in a prokaryotic cell.

2. Chromosomes may be located unrestrained with the prokaryotic cell or within a nuclear membrane in the eukaryotic cell.

3. Chromosomes may be found as single copies per cell (haploid), or in multiple copies per cell, generally two (diploid). In haploid cells, all functional genes present in the cell can be expressed. In diploid cells, one gene may be dominant over the other so that only the dominant gene of each functional pair is expressed. The nonexpressed gene of the pair is called recessive. Functional recessive genes are expressed only when both copies of the recessive type are present (Figure 2.5). Some cell types in mammals are

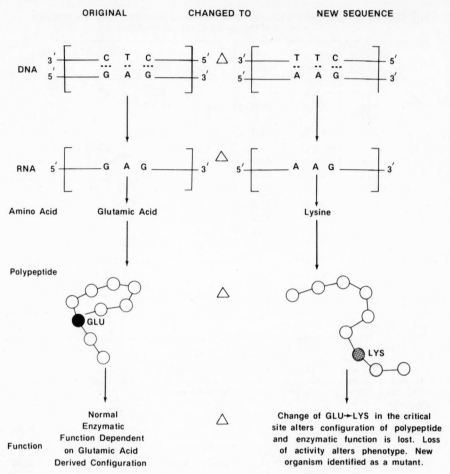

FIGURE 2.2. Gene protein relationships in mutation induction. A change in one codon may produce an amino acid replacement that results in a nonfunctional polypeptide.

found in ploidies other than diploid. Liver cells, for example, are often found to be tetraploid, and gametes are, of course, haploid.

Alternate forms of the preceding characteristics are functions of whether or not the organism is prokaryotic or eukaryotic in nature. Table 2.3 summarizes some of the general traits of prokaryotic and eukaryotic cells as they pertain to the organization of genetic material. It is essential to delineate the similarities and differences between pro- and eukaryotes, since both cell types are routinely employed in genetic testing and some of the differences appear to be critical in specific testing situations (e.g., detection of aneuploidy).

The Cell Cycle and Chromosome Mechanics in Somatic and Germ Cells

The Chromosome

The DNA molecule of a bacterium is not generally referred to as a chromosome, and cell cycle and chromosome mechanics are not particularly relevant with respect to the types of genetic lesions induced in these

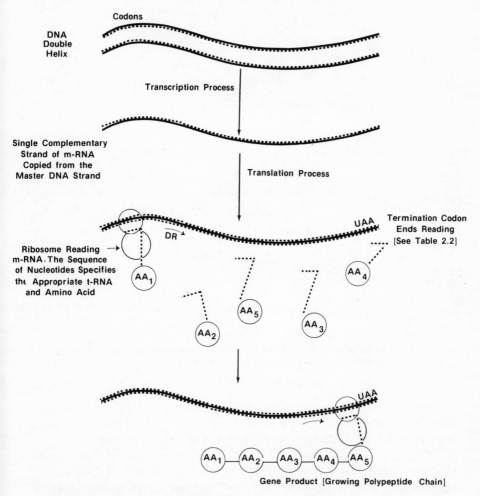

FIGURE 2.3. Essential steps in protein synthesis. Protein synthesis involves two major actions: (a) decoding DNA information into RNA information (transcription) and (b) decoding RNA information into polypeptides (translation). The reference to codon in the top portion of the figure encompasses more than three base pairs and hence more than one codon. m-RNA, messenger RNA; t-RNA, transfer RNA; AA, amino acid; DR, direction of reading; $AA_1 - 5$, amino acids.

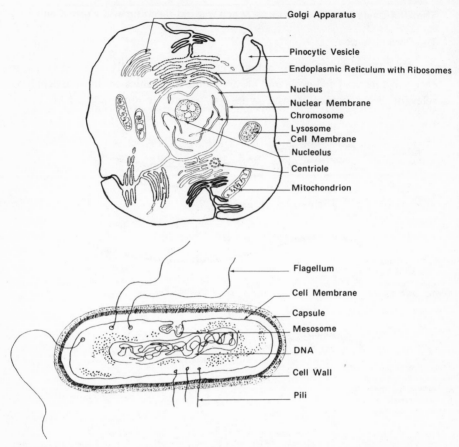

Golgi Apparatus

Pinocytic Vesicle
Endoplasmic Reticulum with Ribosomes
Nucleus
Nuclear Membrane
Chromosome
Lysosome
Cell Membrane
Nucleolus
Centriole
Mitochondrion

Flagellum

Cell Membrane

Capsule

Mesosome

DNA

Cell Wall

Pili

FIGURE 2.4. Prokaryotic and eukaryotic cell types. The two cell types are compared in this figure. Substantial differences are observed in subcellular organization. These illustrations are generalizations; many variations of each cell type exist. Illustrations courtesy of H. Lebowitz.

organisms. However, most test systems in genetic toxicology employ eukaryotic organisms (fungi, insects, plants, mammals), and cell cycle processes become important in conducting the assay and interpreting the results.

Chromosomes consist of DNA containing several hundred to thousands of genes. Complexed with this DNA is a group of basic proteins called histones and some nonhistone proteins.[18] Histones attach preferentially to specific sites on the chromosomes and are considered important in the regulation of gene function since they appear to inhibit the expression (transcription) of the genes to which they are complexed.[5] The union of DNA and proteins provides the basis for the cytologically visible features of chromosomes. This feature is important in cytogenetic assays which rely on microscopic examination of chromosome number and structure. A typical condensed metaphase chromosome illustrating each of these features is shown in Figure 2.6.

Each metaphase chromosome shows a primary constriction called the centromere. The constriction may occur near the middle of the chromosome (metacentric), or near one end (subterminal or acrocentric). The position of the centromere is useful in distinguishing chromosomes, since its location for a given chromosome is constant. Functionally, the centromere is believed to be the attachment site for the spindle apparatus connecting the centriole to the chromosome. This attachment is necessary to separate chromatids during cell division (see Figure 2.9).

A secondary constriction in one or more chromosomes within a set indicates the nucleolar organizer region of the chromosome. The nucleolar organizer region is the site of synthesis on genes for RNA used in ribosome construction. The chromosomal region beyond the secondary nucleolar constriction occasionally appears terminally disassociated from the body of the chromosome, and is referred to as a satellite.[12] The location of these structures is important in chromosome identification and for the visualization of structural alterations induced by chemicals.

Other features that can be localized on the chromosomes are regions of heterochromatin (which has a low content of functional DNA) and euchro-

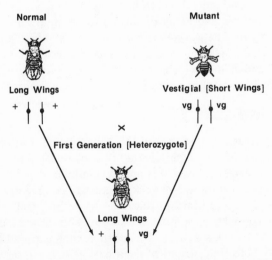

FIGURE 2.5. Dominant and recessive traits. Long-wing trait is dominant and will be expressed either as +/+ or +/vg. Short-wing trait is recessive and will be expressed only when the genotype is vg/vg. This figure also illustrates the relationship between genotype (composition of the gene pair) and phenotype (visable appearance of the organism). For example, two different genotypes (+/+ and +/vg) can both exhibit the same phenotype.

TABLE 2.3
Characteristics of DNA in Prokaryotic and Eukaryotic Cell Types

Prokaryotic	Eukaryotic
Primarily haploid	Primarily diploid
DNA uncomplexed	DNA complexed with proteins forming chromosomes
DNA nonlocalized in the cell cytoplasm	DNA localized primarily within the nucleus of the cell
No morphologic stages in DNA replication	DNA replication described by mitotic cycle consisting of specific cytologic stages
DNA often found as a closed circle	DNA found in linear chromosomes
Replication not associated with cellular organelles	Replication and separation of chromosome associated with cellular organelles called centrioles

matin (which has a high content of functional DNA). The two regions are believed to represent tight and loose coiling regions of DNA, assumed to be the nonactive and active regions, respectively. Staining techniques which rely upon differential protein concentration are referred to as banding techniques, and can be very useful in individual chromosome identification when the overall size and structure appear identical. Each chromosome has its own characteristic and individualized banding profiles.

The total number and array of chromosomes in a cell is referred to as the cell's karyotype. The karyotype for a species is unique and often forms the basis of cellular taxonomy (Figure 2.7). Thorough knowledge of the normal karyotype of a species is essential before entering into cytogenetic studies in which structural or numerical alterations are scored.

The Mitotic Cell Cycle

The cell cycle is oriented toward DNA replication, followed by its division and distribution to each of the two new daughter cells. At the molecular level, DNA replication is an enzymatic process involving a series of enzymes which open the DNA helix, synthesize complementary copies by polymerization of appropriate nitrogenous bases and seal the synthesized strands into a complete new macromolecule. This process is followed by separation of the two new helices. This phenomenon is called semiconservative replication, since one strand of each new helix is conserved and forms the template upon which the new strand is polymerized (Figure 2.8). It is extremely rapid and accurate (less than one error per million bases polymerized).

Eukaryotic cells, especially mammalian cells, have a cell cycle with distinct stages. Although the DNA in eukaryotic chromosomes is believed to replicate in a fashion shown in Figure 2.8, organization and replication of the accessory chromosome constituents requires that these cells undergo a relatively elaborate set of events known as the "cell cycle." Knowledge of these stages is essential in conducting and evaluating various types of tests in genetic toxicology, particularly those which measure unscheduled DNA synthesis (UDS), sister chromatid exchange (SCE), and various *in vitro* and *in vivo* chromosome abnormalities.

Table 2.4 illustrates the various stages in the cell cycle and defines the processes involving DNA/chromosomes which occur in each stage. Mitosis

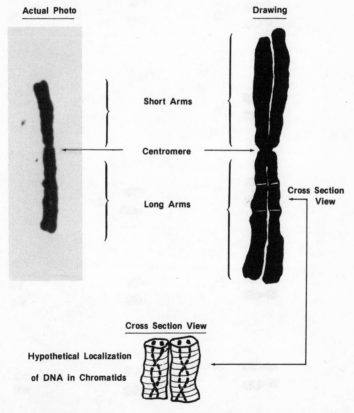

FIGURE 2.6. A condensed metaphase chromosome with chromatids. This chromosome consists of DNA and associated proteins. The presence of chromatids indicates that the chromosome is in the metaphase stage of mitosis just prior to division (see Figure 2.9). The enlarged cross section is a simplified schematic of how the DNA is packaged in a chromosome.

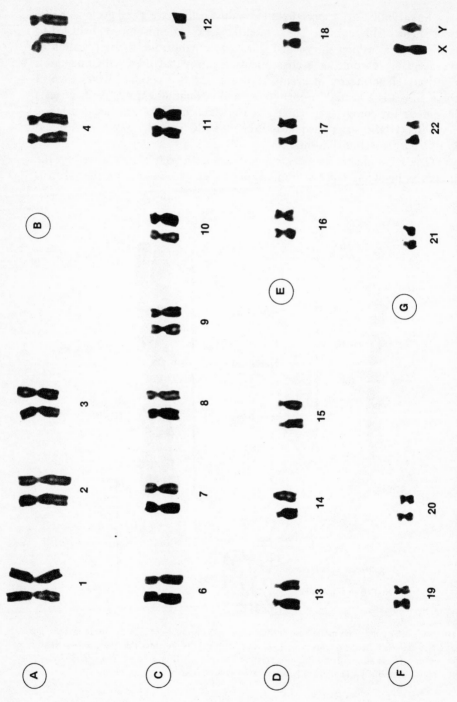

FIGURE 2.7. Human karyotype. This figure illustrates how chromosomes are grouped by size and shape to determine the karyotype of a species. Such karyotypes are used for diagnosis of certain hereditary abnormalities in humans. Photo courtesy of H. Lebowitz.

TABLE 2.4
Stages of a Typical Mammalian Cell[a]

Stage:	G_1	S	G_2	M
Events:	GAP_1	Synthesis of DNA	GAP_2	Mitosis: See Figure 2.9
Relative DNA content:	1	$1 \rightarrow 2$	2	$2 \rightarrow 1$
Common types of aberrations observed[b]:	Chromosome breaks Rings Translocations Deletions Dicentrics Fragments	Mixed	Chromatid breaks Quadriradials Triradials Fragments	
Relative sensitivity to genotoxic effects:	Low	High	Moderate	Low

[a] Shown as relative lengths of the total cell cycle period for *in vitro* conditions.
[b] These aberrations are the types observed when the exposed G_1, S, or G_2 cells are scored at the first metaphase (M1). If the cells are scored at M2 or M3 the aberration types are mixed and lose identity with the cell stage.

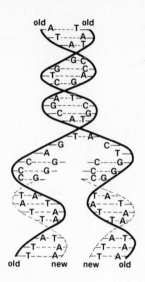

FIGURE 2.8. Structure and duplication of DNA. Semiconservative replication involves the unwinding of the DNA helix exposing each strand. The open strands serve as templates for the enzymatic synthesis of new complimentary strands of DNA. The results will be two duplicate DNA molecules. From Levine, R. P.: *Genetics*, Holt, Rheinhart and Winston, Inc., New York, 1962.

(M)-phase is the stage at which the replicated DNA condenses into discrete chromosomes (chromosomes are not cytologically visible in other stages of the cell cycle), which are then distributed into two new daughter cells. The process of mitosis is arbitrarily divided into five discrete steps (Figure 2.9). Of these steps, metaphase is typically viewed in cytogenetic testing to evaluate chromosomes for breakage and rearrangements. Chemical treatment of cells with colchicine will arrest cells in metaphase for easy viewing. Anaphase has also been used for aberration detection, but the chromosome arrangements are more difficult to interpret and very little use is made of this method in genetic toxicology.[1] Examples of the types of chromosome changes scored in cytogenetic evaluations are shown later in this chapter in Figure 2.16.

Meiosis and Chromosome Mechanics

Meiosis occurs in germinal tissue of diploid species. The end result of this process is the reduction of the diploid (2N) chromosome number to a haploid (N) number, ensuring that each gamete has one copy of each chromosome (and gene) pair. Figure 2.10 illustrates the meiotic process in mammals. This process also includes the mechanisms of independent assortment and distribution of maternal and paternal chromosomes as well as chromosome recombination. See Strickberger[19] for a more detailed discussion of these events.

STAGE	EVENT	CELLULAR CONFIGURATION
Interphase	Chromatin uncondensed in nucleus	
Prophase	Chromatin begins to replicate and condense into visible chromosomes. Nuclear membrane disappears.	
Metaphase	Replicated chromosomes visible as chromatids which arrange for cell division.	*Colchicine will arrest cycle at this pot.*
Anaphase	Chromatids are separated into two daughter cells.	
Telophase	Two identical daughter cells are formed. The nuclear membrane reforms and the chromosomes uncondense and return to interphase appearance.	

FIGURE 2.9. Mitosis. The stages of somatic cell division.

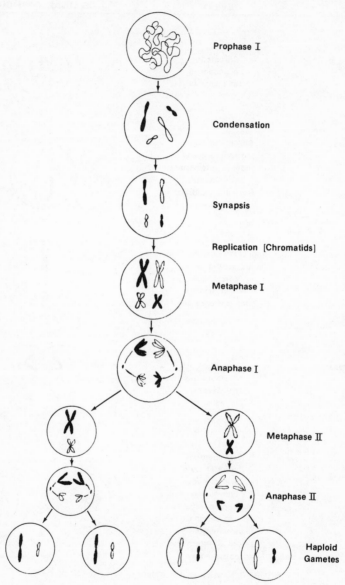

FIGURE 2.10. Meiosis. The stages of germ cell division.

Independent assortment of chromosomes and genetic recombination are two parameters of meiosis important in the randomization of genes before they are packaged as gametes (sperm or ova). Genetic recombination normally occurs during meiosis and involves reciprocal exchanges between chromosomes of a homologous pair adding to the randomization of genes on chromosomes during gametogenesis. Without these processes, genetic variability would be limited to rare mutational events that survive natural selection. While independent assortment and recombination are critical to the understanding of genetic phenomena, they are not essential to most aspects of genetic toxicology testing.

In genetic toxicology, knowledge of the process of meiosis is important when conducting *in vivo* assays such as those employed to detect dominant lethality and heritable translocations. It is essential to know the length of the time required for a chromosome set to pass from the initial stage of meiosis (gonia cells) to the formation of mature sperm or ova, since acute exposure to a mutagenic substance may produce a stage-specific effect. For example, if a test substance is genetically active only on chromosomes undergoing the meiotic Metaphase I cell division, an effect would not be evident unless the exposed male organism is mated to females over a period of time sufficient to permit the damaged cells to traverse through meiosis I and II cell divisions to mature sperm (Figure 2.11). In a typical rodent dominant lethal study using male mice or rats, the results of the stage-specific response described above would not be seen until the 5th or 6th week of mating. Conversely, if the test sample is active only in mature spermatozoa, the response in a dominant lethal test would be evident within the first 2 weeks of mating with little or no effect seen after the 3rd or 4th week since the sperm during the 3rd and 4th weeks would have been meiotic cells at the time of chemical exposure and refractile to its genotoxic properties (for more details see Chapter 8, p. 172). Similar considerations operate for test methods with *Drosophila*, but generally a period of 8–10 days is sufficient to cover the entire meiotic process in this test species.[21]

Alternative dosing methods have been proposed for dominant lethal and mouse heritable translocation assays which reduce the number of animals required in the study by eliminating the sequential mating procedures routinely employed. The proposed alternatives do not, however, identify stage-specific responses. Practical considerations were responsible for the modifications, since the cell stage on which a mutagen acts becomes academic to the toxicologist concerned primarily with detecting an effect. Under the alternative approach, males are dosed over the entire meiotic process, followed by two mating sequences. The rationale is that the chromosomes which will form mature gametes have been continuously exposed during all stages of their cycle through the meiotic process, and thus any stage-specific alterations will be present and expressed in the matings

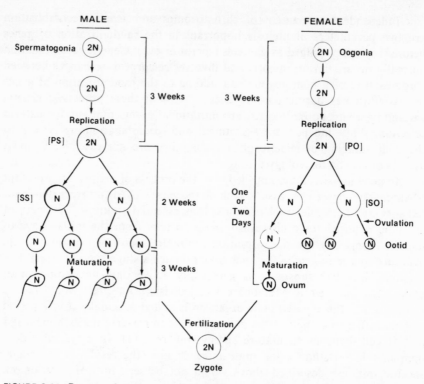

FIGURE 2.11. Passage of gametes through meiosis in rodent species. Male and female stages differ in the number of gametes produced and in temporal sequence. Sperm are produced and cleared continuously. The female germ cell remains as primary oocyte until just before ovulation. Thus, female gametes may experience multiple exposures *in situ* before release. PS/PO, Primary spermatocyte or oocyte; SS/SO, secondary spermatocyte or oocyte.

following continuous dosing. The continuous dosing method is the only practical approach for mouse heritable translocation studies, since sequential weekly matings in this assay would involve prohibitively large numbers of animals. Exposure longer than one complete meiotic cycle is not advantageous since the same pattern of responses would be repeated. The two methods of dosing are compared in Figure 2.12.

Testing which involves exposure of females during the meiotic process is complicated by the fact that, in the female mammal, all ova proceed through meiosis division I and are arrested before meiosis division II. The maturing ova go through division II just before maturation and are released. A few dominant lethal studies have been conducted in female rodents, but the technical difficulties severely limit this type of testing.[3,10] Additional reasons that dominant lethal studies are not performed using females are (1) that oocytes are generally less sensitive to chemicals than sperm and (2) that one cannot easily discriminate between cytotoxic effects impairing ovulation and true genetic effects that result in preimplantation loss. Artifactual preimplantation loss also occurs in male-treated dominant lethal tests but would be expected to be a more reliable indication of genotoxicity than that found in the female-treatment procedure. Good female dominant lethal studies can be performed by taking the time to identify the estrous cycle in the females to be treated and then setting up a regimen in which the treatment can be commenced following ovulation, so that one will be treating cells more sensitive than the oocytes.

DNA ALTERATIONS RESULTING IN GENOTOXIC EFFECTS IN CELLS: MECHANISMS AND CATEGORIZATION

DNA replication is not a perfect process, and in rare instances genetic changes do occur. The occurrence of the changes follows a predictable rate

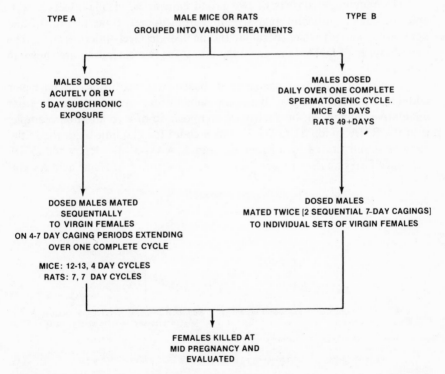

FIGURE 2.12. The dominant lethal assay. Alternate protocols using continuous dosing versus acute exposure. From Brusick, D. J.: Alterations of germ cells leading to mutagenesis and their detection. *Environ. Health Perspectives* **24**:105–112, 1978.

and forms the basis of spontaneous mutation. Spontaneous DNA changes occur at both the nucleotide and chromosome levels. Natural selective processes act on these changes to maintain or eliminate them. Thus, it is not possible to conclude that all mutations are detrimental. Under the appropriate environmental selection pressures, some mutations may be neutral (result in neither beneficial nor deleterious changes) or even beneficial. However, it should be emphasized that very few mutations are beneficial. The genetic composition of human beings has evolved over several million years and exists in a delicate balance. Introduction of random changes, even minor ones, into this carefully evolved system can result in serious diseases, malformations, or death. Although evolution depends on mutation for the introduction of variability into populations, too many changes in too short a time period could seriously reduce the viability of the human species. The consequences of mutaton in humans will be discussed in the next chapter.

A Classification Scheme for Genotoxic Effects

DNA damage consists of two broad categories: visible effects detectable through cytologic analysis of chromosomes (macrolesions), and nonvisible changes which occur at the nucleotide level (microlesions). The specific types of DNA damage falling into these two categories are shown in Figure 2.13.

Microlesions consist primarily of base-pair substitutions or base-pair addition/deletion changes. Base-pair substitution mutations result from qualitative changes in the nucleotide composition of a codon. For example, if in the RNA codon 5', GAG 3', which codes for glutamic acid, the initial guanine is substituted by adenine to form 5' AAG 3', the RNA codon will then specify lysine (See Figure 2.2). If the substitution of lysine acid for glu-

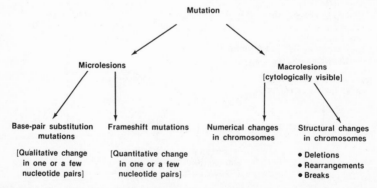

FIGURE 2.13. Classification of the molecular changes in DNA that occur as a result of mutation. From Brusick, D. J.: Alterations of germ cells leading to mutagenesis and their detection. *Environ. Health Perspectives* **24**:105–112, 1978.

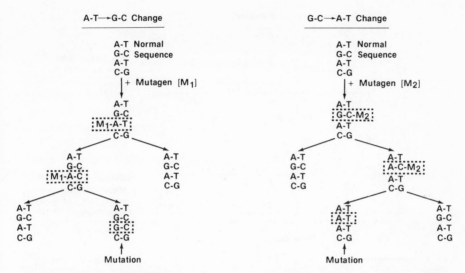

FIGURE 2.14. Mechanisms for base-pair substitution mutation induction. Changes in base sequence brought about by two different mutagens, M1 and M2, are shown. The mutagens differ in their affinities for reaction site. One reacts with purines (A-M1 reaction) and the other with pyrimidines (C-M2 reaction). The proposed mechanisms produce one mutant and one normal cell following induction in what is termed mosaicism.

tamic acid in the polypeptide gene product is incompatible with normal function, the result is a loss in the activity of that gene (probable mutation). Figure 2.14 illustrates a possible mechanism for the induction of this type of base-pair substitution. All combinations of base-pair substitutions are possible. Substitutions of purine for purine and pyrimidine for pyrimidine are designated transitions, while substitutions of purine for pyrimidine or pyrimidine for purine are designated transversions.[9]

Base-pair addition/deletion mutations, also called frameshift mutations, result from the addition or deletion of one or a few nucleotide pairs from the nucleotide complement in a gene.[2] Since the codon sequence of a gene is nonpunctuated, the loss or gain of a single base pair changes the reading frame of the gene; hence, frameshift mutation. This type of change is illustrated in Figure 2.15.

Base-pair substitutions and frameshift mutations are induced by distinctly different mechanisms and generally by distinctly different classes of chemical mutagens. Both types of changes are considered important toxicologically, and like stage specificity in the dominant lethal assay their distinction is of greater academic than toxicologic interest.

Macrolesions can be subdivided into changes in chromosome number (gain or loss of single chromosomes or sets of chromosomes) and changes in chromosome structure (breaks, deletions, rearrangements). Each specific

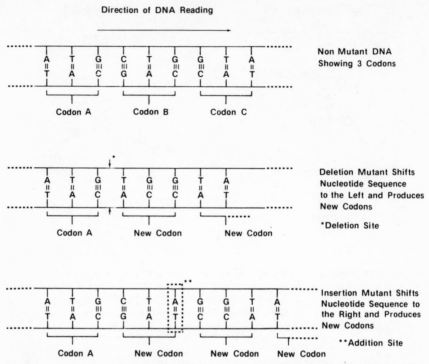

FIGURE 2.15. Mechanisms for frameshift mutation induction. The production of altered codon sequences by insertion or deletion of base pairs will lead to changes in amino acid sequences in the gene product and probably result in defective gene activity.

type of chromosome change has a characteristic designation so that a reasonably high degree of uniformity can be maintained in scoring for these changes. Most cytogenetic studies consist of evaluating mitotic metaphase chromosomes. A glossary of terms and descriptions used in metaphase analysis is provided in Table 2.5.

Variations in chromosome number can result from incomplete dissociation of single or entire sets of chromosomes at metaphase. This may result in cell aneuploidy, monosomy, trisomy, etc., causing serious effects in intact organisms. Some relatively common human disorders such as Mongolism (Down's syndrome) result from variations in chromosome number.

Classically, chromosome aberrations involve structural changes classified according to their time of formation relative to the cell cycle (Table 2.4). Chromatid breaks exhibit one normal and one broken chromatid of a pair, whereas in chromosome breaks both chromatids are broken at identical locations. The same type of relationship is involved in chromatid and chromosome rearrangements. The production of either chro-

TABLE 2.5
Definitions of Abberrations

Term	Definition	Example (in Figure 2.16)
Chromatid gap (tg)	An achromatic region in one chromatid, the size of which is equal to or smaller than the width of the chromatid	f, j
Chromatid break (tb)	An achromatic region in one chromatid larger than the width of the chromatid. It may either be aligned or unaligned	c, e, f
Chromosome gap (sg)	Same as tg, only in both chromatids	a, f
Chromosome break (sb)	Same as tb, only in both chromatids	Not shown
Chromatid deletion (td)	Deleted material at the end of one chromatid	c, g
Fragment (f)	A single chromatid without an evident centromere	c
Acentric fragment (af)	Two aligned (parallel) chromatids without an evident centromere	c, e, h
Translocation (t)	Obvious transfer of material between two or more chromosomes	Not shown
Triradial (tr)	An abnormal arrangement of paired chromatids resulting in a triarmed configuration	g
Quadriradial (qr)	An abnormal arrangement of paired chromatids resulting in a four-armed configuration	e, h
Pulverized chromosome (pu)	A spread containing one fragmented or pulverized chromosome	Not shown
Pulverized chromosomes (pu⁺)	A spread containing two or more fragmented or pulverized chromosomes, but with some intact chromosomes still remaining	Not shown
Pulverized cell (puc)	A cell in which all the chromosomes are totally fragmented	a
Complex rearrangement (cr)	An abnormal translocation figure which involves many chromosomes and is the result of several breaks and mispaired chromatids	c, j
Ring (r)	A chromosome which is a result of telomeric deletions at both ends of the chromosome and the subsequent joining of the ends of the two chromosome arms	i
Minute (min)	A small chromosome which contains a centromere and does not belong in the karyotype	d, j

(Continued)

TABLE 2.5 (*Continued*)

Term	Definition	Example (in Figure 2.16)
Polyploid (pp) or endoreduplication	A cell in which the chromosome number is an even multiple of the haploid number, or N, and is greater than 2N	b
Hyperdiploid (h)	A cell in which the chromosome number is greater than 2N + 1 but is not an even multiple of N	Not shown
Dicentric (d)	A chromosome containing two centromeres	d

matid and/or chromosome aberrations by an agent depends on the nature of the clastogen (chromosome-breaking agent) and the cell cycle stage the target cell was in at the time of exposure. The majority of chromosome-type effects are described from lesions induced in G_1. G_2 exposure generally results in chromatid aberrations. However, most chemical mutagens induce chromatid-type aberrations independent of the cell-cycle stage, provided the cells are examined in the first mitosis after treatment (M1). In cytogenetic analyses, exposed cells can be collected at several time intervals following an acute exposure. Protocols should be structured to ensure that the number of cell cycles that have occurred post treatment are determined and correlated to the scoring process.

Cytogenetic evaluations cannot be used alone to define mutagens, since not all mutagens produce DNA lesions which lead to the formation of chromosome aberrations. Mutagens which act via certain types of base substitution reactions (base analogs, hydroxylamine, and some monofunctional alkylating agents) or those that are pure DNA intercallating frameshift mutagens (acridine and anthraquinone dyes) are not effective clastogens. Bifunctional or polyfunctional alkylating agents, which are good inter- and intrastrand cross-linking agents [nitrogen mustard and triethylenemelamine (TEM)], are the most effective type of clastogenic agents known.

A cytogenetic technique currently in wide use is the analysis for SCEs. This phenomenon was originally studied by Taylor in 1957, but analysis for SCEs on a routine basis only became possible following the development of simple staining techniques that differentiate sister chromatids.[20,23] The mechanisms proposed to explain the staining technique and formation of SCEs are outlined in Figure 2.17. SCE analysis appears to be a very good screeening tool, but the biological significance of SCEs as a genotoxic lesion is not known at the present time.

Another screening technique developed to assess the induction of chromosome damage is the test for production of micronuclei. Schmid[14] and Heddle[7] developed this test independently with a minor difference in the preparation of the slides as the distinguishing feature. The theoretical basis for micronuclei rests with the hypothesis that broken chromosomes or chromatid fragments may lag behind intact chromosomes during the anaphase step of mitosis. During telophase, daughter nuclei are formed. If the broken and lagging chromatin is not included in the main nucleus during telophase, micronuclei are formed in the cytoplasm (Figure 2.18). Thus, it is believed that the frequency of cells containing micronuclei following chemical treatment is indicative of clastogenic activity.[11] While micronuclei may be the results of broken chromosomes or chromatids which produce lagging anaphase fragments, it is reasonable to assume that much of the chromatid damage not expressed after a single cell cycle (e.g., chromatid breaks) will be missed in the micronucleus test using current protocols and that relatively high dose levels may be required to produce effects that could be detected by conventional chromosome analysis at lower doses.

Both macrolesions and microlesions contribute to the overall genetic burden in the human population, and it is difficult, if not impossible, to give greater importance to one class over the other. A comprehensive evaluation of a test material should include provisions for detecting both types of alterations.

Besides the tests which specifically measure microlesions (also referred to as point mutations or specific locus gene mutations) and macrolesions, a diverse group of ancillary tests has been developed to measure other types of genotoxicity. This mixed group of systems is generally grouped under the category of tests for primary DNA damage. Among these tests are those measuring the DNA repair processes, systems detecting mitotic recombination or mitotic gene conversion, and tests measuring spermhead abnormalities in mice. Table 2.6 summarizes several common genetic assays according to the three basic categories and identifies a fourth category specifically addressing oncogenicity.

Repair of DNA Damage

The various systems measuring some parameter of DNA repair[13,16,22] are the most direct measurement of primary DNA damage. Since all normal organisms are capable of some type of repair following chemical insult, a stimulation in the level of repair activity following chemical treatment at sublethal concentrations is a good general indicator that the test sample has DNA-directed toxicity.

To maintain the fidelity and integrity of genetic information, several

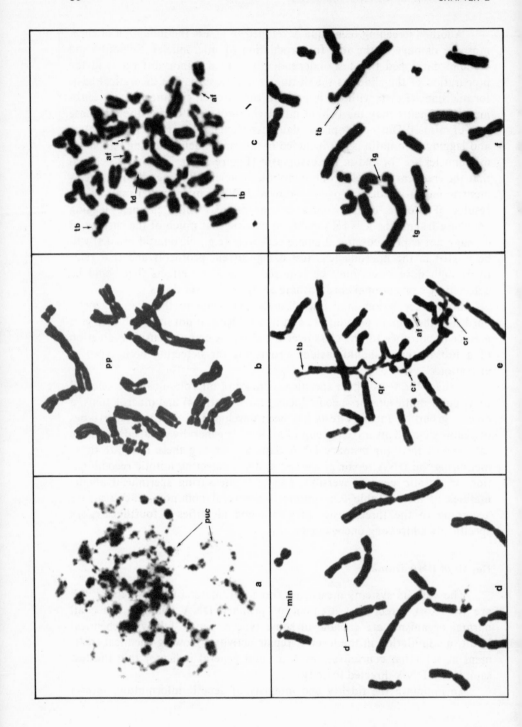

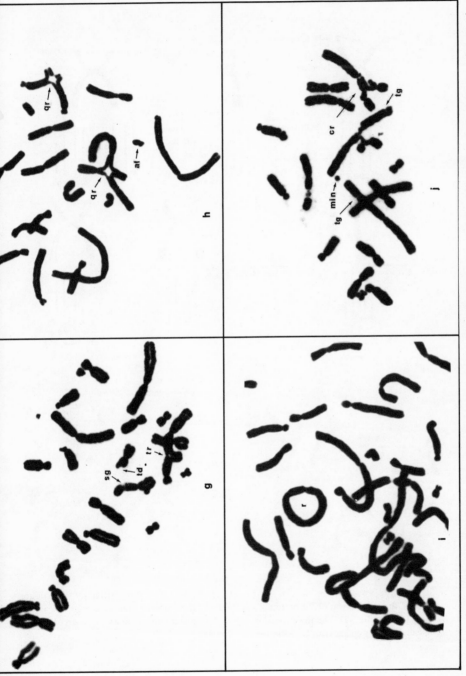

FIGURE 2.16. Examples of typical chromosome abberations scored in mammalian somatic cells. The description and alternation symbol are given with reference to a metaphase spread where the alteration can be visualized.

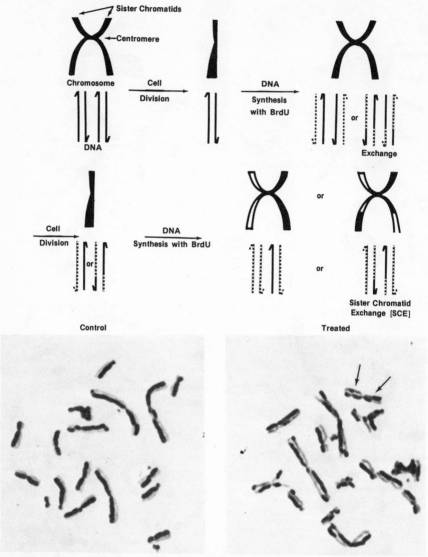

FIGURE 2.17. Visualization of sister chromatic exchange. The top portion of this figure illustrates the mechanism of formation of the light and dark stained areas of the chromosomes (courtesy of D. Stetka). The lower portion shows the stained chromosomes as observed microscopically. Photos courtesy of S. Galloway.

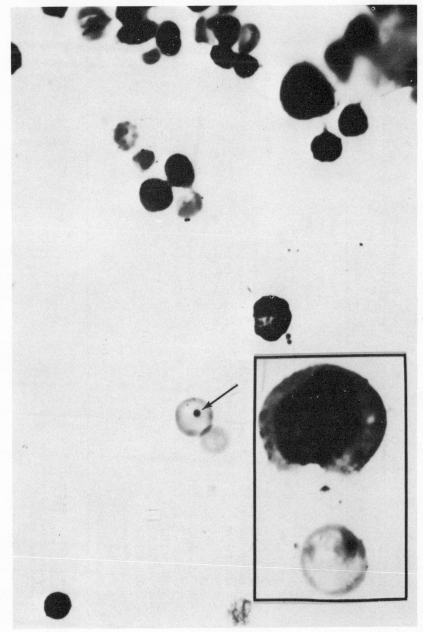

FIGURE 2.18. Micronuclei observed in mouse polychromatic erythrocytes. An enlarged view is also shown in inset. The micronucleus is shown as a darkly stained body in the cell. Photo courtesy of H. Lebowitz.

TABLE 2.6

Basic Categories of Genetic Assays and Some Examples

Gene mutations	Chromosome aberrations	Primary DNA damage	Oncogenic transformation
Bacteria forward and reverse mutation	*In vitro* cytogenetics	*E. coli*: $polA^+/polA^-$ assay	*In vitro* cell transformation:
Ames/*Salmonella*	Mouse bone marrow	*Bacillus subtilis*: $rec+/$	BALB/c 3T3 cells
E. coli: WP_2 $uvrA-$	Mouse micronucleus	$rec-$ assay	SHE^a cells
E. coli: K12 forward mutation	*Drosophila*: X and Y loss	Yeast: D_3, D_4, D_5, or D_7 somatic recombination	Mouse 10T½ cells
Yeast forward and reverse mutation	*Drosophila*: dominant lethal	UDS	
In vitro mammalian cell forward mutation	*Drosophila*: heritable translocation	WI-38 cells	
Mouse lymphoma TK^b assay	Mouse: dominant lethal	Rat primary hepatocytes	
CHO^c HGPRT^d assay	Rat: dominant lethal	SCE analysis	
V79 HGPRT assay	Mouse: heritable translocation		
Drosophila: SLRL^e			
Mouse: somatic mutation test (spot test)			
Specific locus assay in mice			

^{a} Syrian hamster embryo.
^{b} Thymidine kinase.
^{c} Chinese hamster ovary.
^{d} Hypoxanthine quanine phosphoribosyl transpose gene.
^{e} Sex-linked recessive lethal.

types of enzymatic DNA repair processes developed during evolution. DNA is the only molecule in living organisms with a capacity for self-repair. The common feature of repair processes is the ability to remove and replace damaged DNA. Therefore, if a DNA lesion induced by a mutagen can be repaired prior to fixation or stabilization, the net effect may be nil. This is especially true following low-level exposure where excision repair enzymes are not fully saturated by excessive numbers of damaged DNA sites.

Four types of repair mechanisms have been identified. The general characteristics of each is summarized in Table 2.7. Excision repair and, to a certain extent, postreplication repair, appear to be the two processes involved with elimination of most chemically induced DNA lesions, and both contribute greatly to the net genetic effect following exposure of organisms to various genotoxic agents. It has also been anticipated that inherent variation in repair capacity between organisms can lead to differential susceptibility to genotoxic effects. Hart and Setlow reported on the correlation between DNA excision repair capability and lifespan in several mammalian species[6]; there appears to be increased repair in species with longer life-spans (Figure 2.19). This relationship could well play an important role in the extrapolation of genetic effects, since most risk assessment models employ a conservative approach and assume a similar response for both rodents and humans.

In the context of genetic toxicology, repair processes play a critical role at low dose levels. Because of the data shown in Figure 2.19, serious consideration should be given to determining the modifying effect that DNA repair systems have in the assessment and extrapolation of genotoxic

FIGURE 2.19. Relationship of life-span and excision repair capability. Data developed from several mammalian species suggest that animals with longer life-spans have more efficient DNA repair systems than animals with short life-spans. From Hart and Setlow.[6]

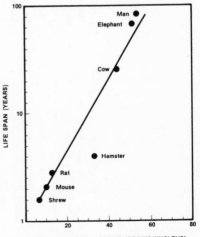

LIFE SPAN (YEARS)

RELATIVE EXCISION REPAIR (GRAINS/NUCLEUS) IN YOUNG FIBROBLASTS EXPOSED TO 10 Jm^{-2} of Uv LIGHT

TABLE 2.7

A Summary of Four DNA Repair Mechanisms

Excision repair	Postreplication repair	Photoreactivation	SOS repair
Damaged DNA areas excised and new sequences resynthesized in gaps	Only occurs during DNA replication	Specific for the repair of ultraviolet light-induced damage (pyrimidine dimerization)	Appears to be confined to prokaryotic organisms such as bacteria; has little or no effect in human DNA repair
Multiple enzymatic functions involved in repair, including recognition of damaged sites, nuclease base removal, polymerase base replacement, and ligase reattachment of DNA ends	Enzymatic resynthesis is involved	Enzymatic action in the presence of visible light opens the dimers to monomers	The repair system is not normally present, but is induced by DNA damage that cannot be repaired by the normal excision or postreplication repair processes
Involved in human DNA repair	Involved in human DNA repair activity	Not easily demonstrated in human cells	Often leads to error-prone repair among surviving cells
Broad range of DNA lesions repaired, including altered bases, larger lesions, or cross-linking of DNA	Repair involves lesions which result in replication gaps such as altered bases or larger single-stranded lesions	Little or no relationship to repair of chemical damage	
	May involve recombinational events		

risk.[6,15] Since the primary repair mechanisms of excision/resynthesis and postreplication are enzymatic in nature, their effect on restoration of non-damaged DNA can be circumvented at high dose levels where the enzyme systems are saturated by DNA lesions and many of the lesions are fixed as mutations or result in chromosome breakage. Extrapolation of DNA damage measured at high dose levels to that which might be induced at very low dose levels must correct for this phenomenon and should consider the impact of DNA repair at low doses.

A complicating factor which alters the apparent benefits is that some repair, particularly the SOS type, may actually lead to mutation (error-prone repair) due to faulty repair synthesis. Until repair-induced mutation can be demonstrated either to function or not function in mammals, the initiation of DNA lesions of all classes must be considered as having potential to lead to heritable damage.

REFERENCES

1. Ad Hoc Committee of The Environmental Mutagen Society and The Institute for Medical Research: Chromosome methodologies in mutation testing. *Toxicol. Appl. Pharmacol.* **22**:269–275, 1972.
2. Ames, B. N., and Whitfield, H. J.: Frameshift mutagenesis in *Salmonella. Cold Spring Harbor Symp. Quant. Biol.* **31**:189–201, 1966.
3. Badr, F. M., and Badr, R. S.: Studies on the mutagenic effect of contraceptive drugs. I. Induction of dominant lethal mutations in female mice. *Mutat. Res.* **26**:529, 1974.
4. Benzer, S.: Fine structure of a genetic region in bacteriophage. In *Papers on Genetics: A Book of Readings* (Louis Levine, ed.), C. V. Mosby, St. Louis, pp. 287–294, 1971.
5. Cattanach, B. M.: Control of chromosome inactivation. In *Annual Review of Genetics*, Vol. 9 (H. L. Roman, ed.), Annual Reviews, Palo Alto, Calif., pp. 1–18, 1975.
6. Hart, R. W., and Setlow, R. B.: Correlation between deoxyribonucleic acid excision-repair and life-span in a number of mammalian species. *Proc. Natl. Acad. Sci. U.S.A.* **71**(6):2169–2173, 1974.
7. Heddle, J.: A rapid *in vitro* test for chromosomal damage. *Mutat. Res.* **18**:187, 1973.
8. Kleinhofs, A., and Behki, R.: Prospects for plant genome modification by nonconventional methods. In *Annual Review of Genetics*, Vol. 11 (H. L. Roman, ed.), Annual Reviews, Palo Alto, Calif., pp. 79–101, 1977.
9. Kreig, D. R.: Specificity of chemical mutagenesis. In *Progress in Nucleic Acid Research*, Vol. 2 (J. N. Davidson and W. E. Cohn, eds.), Academic Press, New York, pp. 125–68, 1963.
10. Machemer, L., and Lorke, D.: Experiences with the dominant lethal test in female mice: effects of alkylating agents and artificial sweeteners on pre-ovulatory oocyte stages. *Mutat. Res.* **29**:209, 1975.
11. Matter, B. E., and Grauwiler, J.: Micronuclei in mouse bone marrow cells. A simple *in vivo* model for the evaluation of drug induced chromosomal aberrations. *Mutat. Res.* **23**:239–249, 1974.
12. Ris, H.: Chromosome structure. In *Chemical Basis of Heredity* (W. D. McElroy and B. Glass, eds.), The John Hopkins Press, Baltimore, 1957.

13. San, R. H. C., and Stich, H. F.: DNA repair synthesis of cultured human cells as a rapid bioassay for chemical carcinogens. *Int. J. Cancer* **16**:284–291, 1975.

14. Schmid, W.: Chemical mutagen testing on *in vivo* somatic mammalian cells. *Agents Actions* **3**:77–85, 1973.

15. Setlow, R. B.: Repair deficient human disorders and cancer. Nature (*London*) **271**:713–717, 1978.

16. Slater, E. E., Anderson, M. D., and Rosenkranz, H. S.: Rapid detection of mutagens and carcinogens. *Cancer Res.* **31**:970–73, 1971.

17. Starlinger, P.: DNA rearrangements in procaryotes. In *Annual Review of Genetics*, Vol. 11 (H. L. Roman, ed.), Annual Reviews, Palo Alto, Calif., pp. 103–26, 1977.

18. Stellwagen, R. H., and Cole, R. D.: Chromosomal proteins. *Ann. Rev. Biochem.* **38**:951–90, 1969.

19. Strickberger, M. W.: *Genetics*. Macmillan, New York, 1968.

20. Taylor, J. H.: Sister chromatid exchanges in tritium-labeled chromosomes. *Genetics* **43**:515–529, 1958.

21. Vogel, E., and Sobels, F. H.: The function of *Drosophila* in genetic toxicology testing. In *Chemical Mutagens: Principles and Methods for Their Detection*, Vol. 4 (A. Hollaender, ed.), Plenum Press, New York, Chapter 38, pp. 93–142, 1976.

22. Williams, G. M.: The detection of chemical carcinogens by unscheduled DNA synthesis in rat liver primary cell cultures. *Cancer Res.* **37**:1845–1851, 1977.

23. Wolff, S., and Perry, P.: Differential Giemsa staining of sister chromatids and the study of sister chromatid exchange without autoradiography. *Chromosoma* **48**:341–353, 1974.

24. Yanofsky, C., Drapeau, G. R., Guest, J. R., and Carlton, B. C.: The complete amino acid sequence of the tryptophan synthetase A protein (α subunit) and its colinear relationship with the genetic map of the A gene. In *Papers on Genetics: A Book of Readings* (Louis Levine, ed.), C. V. Mosby, St. Louis, pp. 335–37, 1971.

CHAPTER 3

The Consequences
of Genotoxic Effects in
Humans and Other Mammals

INTRODUCTION

The major factors responsible for the concern over genetic effects can be divided into two areas (Figure 3.1). The first is the concern for the protection of the human gene pool. This forcing factor may be the most significant aspect of genetic testing, but it is the least appreciated by most nongeneticists involved with safety evaluation because of the difficulty of demonstrating induced mutation in humans. The second area is that of oncogenesis. The intimate relationship between the tumorigenic and genotoxic properties of chemicals makes genetic testing an attractive screening and prioritizing technique for chemicals of unknown oncogenic potential. This second forcing factor has been the primary driving force behind the rapid expansion of genetic toxicology as a discipline.

Another factor of importance in assessing the consequences of genetic effects is the location of the affected cell (Figure 3.2). If mutations occur in cells which are not part of the reproductive system (somatic cells), the resultant alteration affects only the individual exposed to the genotoxic agent and will not be transmitted. If the alteration occurs in gametes (sperm or ova) or the stem cells which give rise to gametes, the alteration may affect subsequent generations. This latter circumstance is the primary reason that it is difficult to demonstrate a cause–effect relationship in humans between exposure to a mutagen and the production of altered individuals. The mutation induced in the gametes of an exposed individual may be expressed

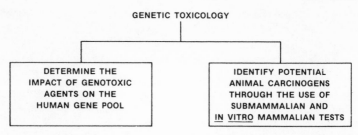

Figure 3.1. The two roles of genetic toxicology which have been forcing factors in its development as a discipline.

immediately if it is dominant, or it may not be expressed for several generations if it is recessive. Human epidemiological studies are of limited value in developing a case for restriction of mutagens because, in most cases, the cause-and-effect portions of the relationship are separated by a considerable length of time. Concern over genotoxicity must be developed through an awareness of the serious consequences likely to occur 100 years or more from today if human exposure to significant levels of mutagenic agents is not prevented.

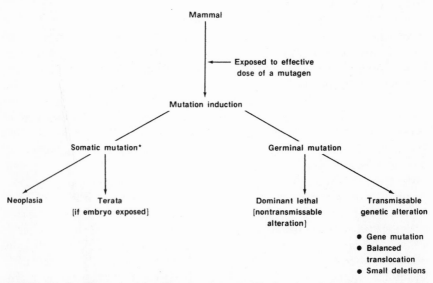

FIGURE 3.2. Possible mutagenic effects in mammals. The location of the mutagenic event is important in determining whether its effect will have an impact on the exposed individual or individuals in subsequent generations. Somatic cell alterations can be detrimental to exposed individuals, whereas the alterations induced in germ cells will affect subsequent generations. * Also implicated in certain types of heart disease.[4] From Brusick, D. J.: Alterations of germ cells leading to mutagenesis and their detection. *Environ. Health Perspectives* **24**:105–112, 1979.

Although the relationship between mutagenesis and carcinogenesis has been intensively studied and reviewed, it is preferable to discuss these two areas separately, since the approaches and goals of testing for animal mutagens and carcinogens are distinctly different in many respects.

GENE POOL CONSEQUENCES

The animal species of ultimate concern in toxicologic investigations is the human one, and in particular of those individuals in the reproductive portion of their life span. The human gene pool is the sum total of the genes available in the reproductive cells of the population that will be transmitted to the next generation. All of the genes which form the present generation of humans were acquired from the previous generation's gene pool, and the genes that we pass to our offspring will form the gene pool of the next generation. There are obviously deleterious genes among the total number of genes within a given population, and their frequency in random mating populations is to a great extent related to their frequency in preceding generations. Transmission of specific mutations is followed through several generations using pedigree charts of the type shown in Figure 3.3. The total estimated level or frequency of these deleterious genes is designated as the genetic load of a population, and may be thought of as a genetic burden transmitted from generation to generation. The specific origin of this burden is not known; however, the processes of mutation (induced and spontaneous) and natural selection (preferential persistence of a given gene in random mating populations) are important processes contributing to the genetic load of a population.[24] New mutations increase the genetic load (burden), and a large-scale introduction of new mutations may eventually result in a generalized reduction in the reproductive capacity and overall viability of a species. The result of a high genetic load in a population is termed genetic death and may be expressed in high levels of prereproductive death or sterility.

It is often asked if all mutations are deleterious, since without mutation new species would not have emerged during evolution. While it is true that mutations provide the substrate for evolutionary change, change in and of itself is not necessarily beneficial. Evolutionary development involves both time and variation. The environmental scene changes with time, and biological variation results from new gene combinations formed during gametogenesis. If too much variation occurs in a short time span, or if too much time elapses without genetic change, serious biological problems may arise for a species.

Evolution has produced a complex of genes in humans which contains information to produce a most sophisticated creature, and random change in

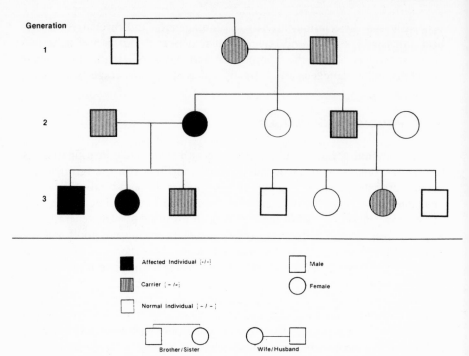

FIGURE 3.3. Pedigree analysis of an autosomal recessive genetic disease traced over three generations. In the first generation none of the individuals are affected although two are heterozygous carriers. The disease is expressed in one of the daughters of the two carriers. In the third generation two individuals are affected with the disease. Construction of a family pedigree can be used to define the type of inheritance and therefore the nature of the genetic lesion.

the composition of this gene set in the present environment is likely to result in genetic dysfunction leading to disease or death. The same mutation which might reduce the viability of a species living under one set of environmental conditions might result in an advantageous trait under other circumstances. Therefore, discussions of beneficial or deleterious mutations are relative and can only be interpreted when the natural selection process is fully understood. Therefore, while a low mutation rate may be essential for the survival of the human species, a rapid increase in mutation rate resulting from exposure to environmental mutagens will almost certainly result in an increased genetic burden for the gene pool and a corresponding increase in health care costs associated with genetic dysfunction in subsequent generations. For example, it is estimated that direct damage to the genetic material or induced modulation of its function could be the cause for a significant portion of the human health burden.[1,9] The goal of genetic toxicology is to protect the integrity of the gene pool inherited from our ancestors and to pass to our heirs a healthy set of gametes.

THE DIRECT EFFECT OF MUTAGENS ON THE TRANSMISSIBLE GENE POOL

One might conclude that with a normal spontaneous mutation incidence and chemical-induced mutations occurring from environmental exposure that the human gene pool would rapidly deteriorate. However, many new mutations induced by genotoxic agents are probably not transmitted via gametes for a variety of reasons (Table 3.1), and even if a mutation is fixed, its chance of survival in the gene pool is low. For example, if a particular normal gene is

TABLE 3.1

Factors Involved in Preventing the Establishment of a Single New Mutation in the Gene Pool[a,b]

Critical step	Reason for loss of the mutation
Meiosis[c]	Mutations induced in stem or gonial cells must be able to traverse events in meiosis such as DNA/chromosome replication. Many types of chromosome aberrations are lethal when subjected to the processes of mitosis and cell division.
Ejaculation	Assuming that the mutation is compatible with meiosis and resides in a mature sperm, it may be lost in the seminal fluid and never reach the region of fertilization.
Fertilization	From the millions of cells in an ejaculation, the number of sperm involved in fertilization is low (1–20, depending on the species). Thus, most sperm from an ejaculation will be lost.
Implantation	Mutations which survive meiosis and reside in a sperm which fertilizes an egg may be lost if their expression interferes with zygote implantation or embryonic development. Early death and resorption will occur. This would probably be true only for dominant mutations. Postfertilization repair, known to occur in mammalian zygotes, could affect expression of mutation.
Embryonic development	Mutations may affect processes of cellular differentiation or hormonal regulation necessary for maintenance of the embryo and fetus. Expression of these mutations may result in early abortion.
Early death	Mutation which affects the prereproductive survival of the progeny will be eliminated by early death of the progeny before transmission and fixation of the mutation in the gene pool.
Transmission capacity	Assuming that an F_1 individual is heterozygous $(+/-)$ and mates with a normal $(+/-)$ individual, the probability of fixation of the new mutation is a function of the number of their offspring. Sterility is also a factor.

[a] This logic applies to a single new mutation induced in a premeiotic or postmeiotic cell. It does not consider the total number of mutations per cell or total number of mutations per cell populations.

[b] The events shown in this table make it appear that virtually no new mutations would be transmitted from one generation to the next. However, the events were developed by looking at a single mutation in a single maturing gamete. When an animal is exposed to a mutagen, stem cells and all developing gametes and all genes within the gametes are potential targets. Thus, for a potent mutagen it may be possible that at least one new mutation will be induced in each maturing gamete, ensuring that some genetic change will be transmitted to the progeny.

[c] Applicable only to mutations in premeiotic cells. All remaining steps applicable to either pre- or postmeiotic cells.

identified by *T*, then most individuals would be represented as *TT* (homozygous). A mutation at the *T* gene (represented by *t*) in a gamete would result in an individual represented as *Tt* (a heterozygote carrier). Since most individuals in the population would be *TT*, mating between the newly formed mutant carrier and a normal individual (*Tt* × *TT*) would result in a 50% probability of producing a *Tt* heterozygote. Thus, for a family with only one child the probability that a new mutant *t* gene will not be transmitted is one-half, if there are two children the probability of nontransmission is only one-quarter, and as the number of children grows the probability of nontransmissibility grows smaller. However, even if the *t* gene is transmitted to one of the children, there is again a high probability that it will not be passed to the next generation. Thus, under normal circumstances it is difficult for a mutant gene to become established in a population of random mating individuals without some sort of selective pressure to maintain its existence. This relationship is independent of whether or not the mutation is dominant or recessive.

The mutation would have a high probability for fixation in a situation where the mutant *t* gene confers some selective advantage to the *Tt* carrier, making this individual more likely to survive and/or mate than the *TT* individual. Evidence for this type of heterozygous selection has been reported for some human diseases. One of the more notable examples is sickle cell disease, in which the unaffected *Ss* heterozygote carrier appears to be more resistant to malaria parasites than normal *SS* individuals. Thus, under the proper set of environmental conditions (i.e., high levels of malaria) the *Ss* carrier will have a greater survival opportunity than the normal individual. However, the individuals affected (*ss*) suffer from sickle cell disease, which can be severely debilitating or lethal.

Genetic alterations fall under several categories (Table 3.2). Among these categories are alterations which result from the gain or loss of chromosomes, portions of chromosomes, or entire sets of chromosomes. The most common chromosome anomalies occur in the sex chromosome pair resulting in either XO (Turner's syndrome), XXY (Klinefelter's syndrome), or XYY individuals. Chromosome anomalies such as deletions or rearrangements also produce characteristic human disorders. Chromosome aberrations are either balanced or unbalanced. Unbalanced events are usually lethal to the cell, whereas balanced events may be compatible with survival.[6] In Table 3.3 it is seen that the frequency of different types of aberrations varies between live births and spontaneous abortions.

The mechanism believed to produce changes in chromosome numbers is nondisjunction, that is, the failure of a chromosome pair to separate during the meiotic divisions (Figure 3.4). Nondisjunction may occur in sex chromosomes and in nonsex chromosomes (autosomes). Down's syndrome (Mongolism) results from the presence of an extra chromosome (number

TABLE 3.2
Examples of Genetic Disorders in Humans

Disorder	Estimated frequency/10^3 population[a]	Typical examples
Chromosomal abnormalities	6.86	Down's syndrome (trisomy 21)
		Klinefelter's syndrome (XXY)
		Turner's syndrome (XO)
		Cri du chat (deletion of chromosome)
		Numerous other trisomies
		XYY
Dominant mutations	1.85–2.64	Familial polyposis (AD[b])
		Neurofibromatosis (AD)
		Huntington's chorea (AD)
		Hepatic porphyria (AD)
		Crouzon's craniofacial dysostosis (AD)
		Achondroplasia dwarfism (AD)
		Retinoblastoma (AD)
		Anitidia (AD)
		Chondrodystrophy (AD)
Recessive mutations	2.23–2.54	Xeroderma pigmentosa (AR[c])
	0.78–1.99	Duchene muscular dystrophy (XR[d])
		Hemophilia (XR)
		Lesch–Nyhan syndrome (XR)
		Sickle cell disease (AR)
		Galactosemia (AR)
		PKU (AR)
		Diabetes mellitus (AR?)
		Fanconi's syndrome (AR)
		Albinism (AR)
		Cystic fibrosis (AR)
Polygenic (complex inheritance)	26.00–32.00	Cleft lip
		Anencephaly
		Spina bifida
		Clubfoot
		Idiopathic epilepsy
		Congenital heart defects

[a] A Consultative Document on Guidelines for the Testing of Chemicals for Mutagenicity. Committee on Mutagenicity of Chemicals in Food, Consumer Products, and the Environment, Department of Health and Social Security, Great Britain, March 1979.
[b] AD = Autosomal dominant.
[c] AR = Autosomal recessive.
[d] XR = X-linked recessive.

TABLE 3.3

Comparison of the Frequency of Types of
Chromosome Abnormalities in Human Live Births and
Spontaneous Abortions[a]

Observed relative frequency	Live births	Spontaneous abortions
Highest	Structural	Trisomy
↓	Monosomy	Monosomy
	Trisomy	Polyploid
Lowest	Polyploid	Structural

[a] This data illustrates the effects of chromosome alterations on organism lethality. Chromosomal changes involving sets or whole chromosomes are generally less compatible with life than breaks and gaps, and are frequently observed in abortions. Structural aberrations seldom lead to spontaneous abortion.

21); that is, individuals with Down's syndrome have three chromosomes rather than two.[15] This slight change in chromosome number produces extensive physiological and psychological changes in the affected individuals, including mental retardation and shortened life expectancy. Autosomal nondisjunction is associated with late pregnancy in women[15]; however, genotoxic agents have also been identified with the production of nondisjunction in submammalian experimental systems.[5] It is currently believed that most nondisjunction in mammals leading to trisomy of autosomes or sex chromosomes is of a spontaneous nature. Considerably more research is needed to establish whether or not chemicals can specifically induce this type of genetic alteration.

Chemicals that preferentially produce chromosome alterations without the concomitant induction of mutations at the gene level are designated clastogens. Most aberrations are lethal to the cell, but certain types such as balanced translocations and small deletions can be transmitted through the germ lines.[6] For instance, the mutations forming the latter three categories in Table 3.2 are gene mutations which have a high level of transmissibility. They affect only a single gene and do not interfere with chromosome replication or division.

Gene mutations are classified in Figure 3.5 into categories according to their expression and impact on the genetic load. The gene pool impact of three types of mutations are illustrated in this figure. A dominant mutation is expressed immediately in the first generation following its production. If the dominant mutation is a lethal one, it will be expressed early in the zygote or developing embryo. The sperm may not be capable of fertilizing the ovum, or if it is successful, the resultant zygote will die before a mature fetus is formed. Dominant lethal mutations, therefore, have little impact on the gene pool and are self-limiting events.

Dominant viable mutations are also expressed immediately, but do not result in embryonic death and can be observed in the first or F_1 generation. These mutations will have an impact on the gene pool since they are transmissible. However, since both their presence and transmission are readily identifiable, a certain level of control can be implemented. An autosomal dominant mutation, G, will be expressed in a heterozygous Gg carrier, and if that carrier mates with a normal gg individual, there is a 50% probability that a child will receive the dominant G gene. Depending on the severity of the effect of the G mutation, the Gg and gg parents will have the benefit of knowing the probability of having a Gg child and can decide in advance if the risk is acceptable. Dominant viable and sex-linked traits can be easily

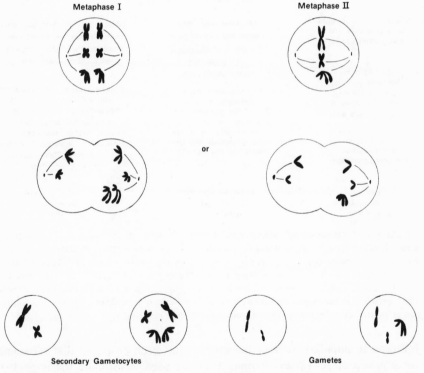

FIGURE 3.4. Illustration of the process of nondisjunction occurring at either metaphase I or metaphase II of meiosis. In both situations shown, one cell receives two chromosomes of the same pair while the sister cell loses an entire chromosome. Nondisjunction need not always affect an entire chromosome, but occasionally small pieces are involved. Nondisjunction generally leads to serious genetic problems in humans. Trisomy 21—mongolism—is a typical example.

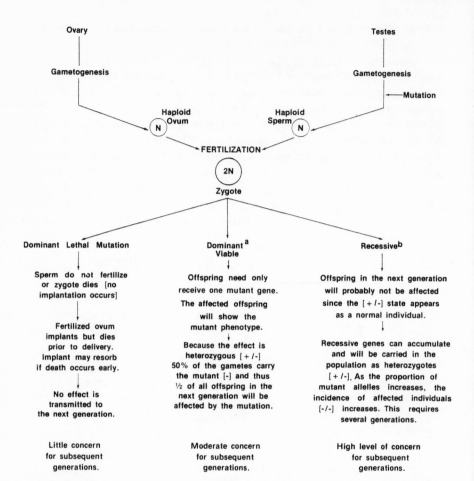

FIGURE 3.5. Consequences of various types of germ cell mutations in mammals. The impact of a mutation varies with the nature of the alteration. [a] Dominant viable mutations may affect a number of biochemical or structural gene functions, giving rise to mutants that have few deleterious effects or to those with near lethal effects. The impact of this gene on the gene pool can be controlled by genetic counseling. [b] Recessive mutations range from neutral to lethal effects. Their presence in the population as (+/−) is almost impossible to detect, and thus no control can be exerted over their transmission from generation to generation.

followed in populations once detected. It has been proposed that screening for a group of 10–12 such human traits be used to estimate the impact of mutagens on exposed populations. However, very large populations would be required for accurate estimates.

Recessive mutations are those events which are not expressed unless the affected individual receives mutant genes from both parents. An example of this process can again be illustrated by sickle cell disease. Two potential carriers (*Ss*) may be unaffected and not realize they carry the *s* mutant gene. Upon mating, the probability of producing an *ss* child is 25%, since *Ss* × *Ss* gives probabilities of 25% for *SS*, 50% for *Ss*, and 25% for *ss*. It may only be after having a child with sickle cell disease that the parents realize they are *Ss* carriers.

Semidominant alleles and X-chromosome (sex-linked) mutant alleles present two additional situations which contribute to the genetic burden. Both of these alterations are expressed more rapidly than recessive mutations, but generally not in the first generation like dominant alleles. Their expression is restricted to specific conditions (e.g., sex of offspring).

New recessive mutations introduced into the gene pool will probably not be expressed for several generations because of the requirement for homozygosity. Rapid accumulation of recessive mutations in the gene pool as a result of exposure to environmental mutagens may result in the eventual deterioration of the genetic viability of the human species. If subsequent generations eventually express the mutants accumulated during current exposures to environmental mutagens, the problem would be more serious than just the costs associated with health care for affected individuals. The mutations induced during the previous exposed generations cannot be removed from the gene pool, because heterozygous carriers appear normal and go undetected. These results, then, are a permanent burden to the species and will continue to show up even though the causative agent inducing the mutations has been removed from the environment.

The most complex situation occurs when hereditary traits are influenced by several recessive gene sets. These traits, called polygenic characters, require a specific combination of genes to result in the expression of a particular trait. Many new mutations could be induced in these genes without ever being noticed.

Assuming that a new mutation is transmitted and established in a population, the expression of the mutant phenotype depends on several factors not well understood. Various secondary factors can influence the level of expression for any mutant gene. The expression may vary among individuals (dependent on genetic background) with sex and age. The frequency with which the trait is expressed is designated penetrance, and the level to which the mutant is expressed is called its expressivity. Thus, mutations with complete expressivity will show the full set of features associated with the alteration, whereas mutants with incomplete expressivity will result in some fraction of the characteristic features in affected individuals. The penetrance and expressivity of a particular mutation cannot be assessed in advance of its expression in a given individual except in a most general way. For example, the severity of a particular genetic disease may range from subclinical to highly debilitating even among members of the same family.

A certain perspective must be maintained when assessing the impact of environmental mutagens on the human gene pool. This perspective should encompass the effect of modern medicine on maintaining deleterious genes in the population. With the aid of newborn monitoring and appropriate therapy, the effects of many genetic diseases which resulted in prereproductive death can now be eliminated. This permits more affected individuals to lead "normal" lives, which includes producing offspring. Without doubt, medical treatment of genetic disorders will contribute more to a higher genetic load via proliferation of preexisting mutations than environmental mutagens via new mutations.

THE RELATIONSHIP OF GENOTOXIC EFFECTS TO OTHER TOXICOLOGIC END POINTS

Because of the fundamental role that genes play in all aspects of living organisms, the concept that altered genes will lead to various disease states has considerable merit.[1] Direct evidence is available for diseases inherited as single-gene traits (Table 3.2). In these examples, mutant genes result in defective mental capacity, sexual characteristics, vision, bone development, skin pigmentation, blood cell formation, muscle development, enzymatic processes, and embryonic development.

Other types of toxicologic end points appear to be under some type of genetic control, but the evidence is only indirect. Among these end points are:

1. Oncogenesis (certain forms such as retinoblastoma and familial polyposis are clearly inherited)[17]
2. Teratogenesis[16]
3. Sterility or semisterility[10]
4. Heart disease[4]
5. Aging[9]

While it is known that genotoxic agents can produce alterations which interfere with cell differentiation and produce terata, a high proportion of teratogenic agents are not mutagens and act through nongenetic mechanisms.[10] Consequently, short-term mutagenicity tests cannot be relied upon to define teratogenic agents. However, *in vitro* embryo culture techniques relying upon the detection of changes in molecular events such as protein synthesis are being explored as potential predictive tests for teratogenicity.

The primary emphasis placed on genetic testing, however, has been its relationship to carcinogenesis. Several factors have been instrumental in the increased emphasis on short-term carcinogenicity tests, primarily the high cost and the long duration required to perform rodent oncogenicity studies. Costs for typical chronic oncogenicity studies in mice and rats have

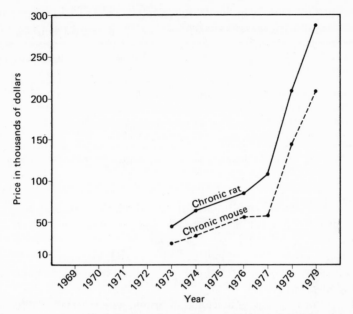

FIGURE 3.6. Price trends 1969–1979 (based on average annual price) for chronic carcinogenicity bioassays. The prices are based on average annual prices. The increases are a result of many factors, including regulatory changes requiring more stringent protocols (GLP), inflation of wages, and increased facilities demand and overhead costs.

increased dramatically, as shown in Figure 3.6, making the decision to perform such studies on a particular chemical a major financial and logistical undertaking. Thus, any information that can be obtained in advance regarding the likely outcome of such studies is extremely valuable in go/no-go decision-making processes.

A somatic mutation theory for the etiology of oncogenicity was proposed by Bauer in 1928.[3] The available data base was insufficient to support such a theory through the 1950s[8] and even the late 1960s.[20] The type of correlative data base to support a mutation step in the induction of tumors was not available until the early 1970s, when sensitive microbial mutation assays were coupled with microsomal metabolic activation (S9 mix) systems.[2] This appears reasonable as many of the experimental animal carcinogens are biotransformed *in vivo* into biologically active agents. Most bacteria or *in vitro* assays cannot carry out such transformation without an exogenous metabolic source. Analysis of animal carcinogens in S9 microsome enzyme-supplemented microbial tests produced highly suggestive evidence that most (85–90%) animal carcinogens were also mutagens[21] (Table 3.4). The degree of correlation between these two phenomena is

TABLE 3.4
Correlation between Animal Carcinogens and Bacterial Mutagens[a]

	Ames et al.	Purchase et al.	Sugimura et al.	Average
Carcinogenic mutagens	89.7%	91.4%	85.0%	88.0%
Carcinogenic nonmutagens	10.3%	8.6%	15.0%	11.95%
Noncarcinogenic nonmutagens	87.0%	94.0%	74.1%	84.5%
Noncarcinogenic mutagens	13.0%	6.5%	26.0%	15.5%

[a] From Nagao et al.[21] All studies employed a bacteria assay coupled with a hepatic S9 activation system.

strongly indicative of an intimate functional relationship, and the possibility that a genotoxic event is essential to the oncogenic process has gained considerable support.[7,18,23]

Several other pieces of indirect experimental evidence have been cited[18] to confirm the functional relationship between the toxic end point of mutation and oncogenicity.

1. Tumors appear to be clonal in origin, which is evidence for a single-cell origin and consistent with somatic mutation. Tumors appear to result from a process of cell transformation followed by cell proliferation. Every cell which undergoes transformation to a malignant state will not necessarily produce a tumor, because of immune surveillance and other *in vivo* control phenomena.

2. Transformed malignant cells have phenotypic properties that are different from their nontransformed precursor cells; this is evidence that a genotypic change has produced the new phenotype.

3. Transformed cells transmit their phenotypic properties to all progeny cells. This is evidence that the control of malignancy is associated with the hereditary components of the altered cell.

4. Transformed cells will grow into a tumor if transplanted into a healthy syngeneic host animal. This demonstrates that the tumorigenic properties are directly associated with the malignant cells.

5. Hart et al. have demonstrated a DNA dependency in ultraviolet-induced neoplasia in the Amazon molly.[14]

6. Cell-fusion experiments between malignant and nonmalignant mouse cells identified a specific gene locus on chromosome 4 to be involved with malignancy.[22]

Based on this indirect evidence, a generalized hypothesis associating mutagenic and carcinogenic mechanisms has been proposed (Figure 3.7). Chemicals which show both properties and fit this scheme have been referred to as mutacarcinogens, DNA carcinogens, or initiating carcinogens, implying that their mechanism of action involves genotoxic events. This hypothesis is supported by additional information derived from mutagen-susceptible organisms. Humans who are deficient in DNA repair are more susceptible to mutagenic agents and also experience a higher susceptibility to cancer, as shown in Table 3.5.[22,23]

There are, of course, numerous problems associated with trying to use indirect evidence as a basis of proof. While the correlation between animal carcinogens and their responses in *in vitro* assays can be shown to be high, the exact correlation value may range from 0.70 to 0.95, depending on the composition of the group of chemicals employed. Loading the group with aromatic amines, polycyclic hydrocarbons, nitrosamines, and direct-acting

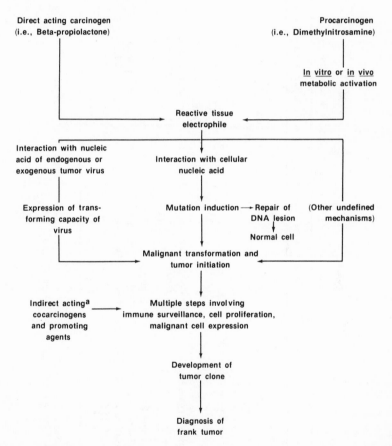

FIGURE 3.7. Proposed relationship between carcinogenicity and mutagenicity. The model shown accounts for many facets of what is known about tumor production in mammals. While there is significant indirect support for this model, formal proof is not yet available. [a] There are a number of agents known to influence the expression of malignancy in cells once the initiation process has occurred. However, these agents are not capable of producing the initiation event.

TABLE 3.5

Repair Deficient Diseases and Cancer Susceptibility[a]

	Xeroderma pigmentosum	Ataxia telangiectasia	Fanconi's anemia
Frequency			
Homozygotes (−/−)	1/300,000	1/40,000	1/300,000
Heterozygotes (+/−)	1/300	1/100	1/300
Cancer probability[b]			
Homozygotes less than age 20	Skin cancer: >0.5 (melanoma: >0.1)	0.1 (lymphoreticular: 0.06; leukemia: 0.02)	>10-fold greater than normal
Heterozygotes	5-fold greater than normal in South	5-fold greater than normal <45 years, 50% greater mortality than average	Like normal
Etiologic agent	Sunlight	?	?
Cell sensitivity	Ultraviolet and mimetics	X-rays, alkylating agents	Cross-linking agents
Repair deficiencies associated with trait	One or more of excision, photoreactivation, and post-replication (≥ 7 groups)	Some cell strains defective in "X-ray" repair (≥3 groups)	Some cell strains defective in cross-link repair

[a] From Setlow, 1980, *Arch. Toxicol.* (in press).
[b] Approximate average cancer probabilities: skin cancer prevalence: 0.005; melanoma incidence: 6×10^{-5}/year; lymphoreticular cancer: 13×10^{-5}/year; leukemia ($t_{max} \approx 4$ years): 42×10^{-6}/year.

alkylating agents, will produce an exaggerated relationship; whereas the inclusion of a high proportion of metal carcinogens, hormones, halogenated organic molecules, and aromatic solvents will result in a reduction of the correlation. Thus, correlations can be manipulated and probably represent the least convincing evidence for the carcinogen/mutagen relationship. Table 3.6 gives the approximate predictive values for several types of *in vitro* and submammalian tests selected from published literature with a relatively balanced grouping of chemicals. Most *in vivo* genetic assays are generally not used to detect oncogenic substances.

The correlations developed by the investigations summarized in Table 3.6 indicate that roughly 90% of the rodent carcinogens will be positive when tested in *in vitro* or submammalian tests. The implication derived from this relationship as applied to genetic toxicology might be that 90% of the mutagens identified by the group of *in vitro* and submammalian tests ought to be rodent carcinogens. This is not likely. In fact, there is no reason to presume, if 90% of all rodent carcinogens are genetically active, that 90% of all mutagens will be rodent carcinogens (Figure 3.8). In addition to mutagenic carcinogens, there will be mutagenic noncarcinogens (i.e., sodium azide and Dichlorvos) based on current test methods. Mutagenic noncarcinogens are usually identified in the literature as false positive. But this may not be the case, since the tests results from negative animal studies may be subject to change as animal testing procedures are modified and improved. An example of such a change is the Japanese food additive AF-2, which was initially designated noncarcinogenic in rodents based on chronic

TABLE 3.6
Screening Tests

Assay	Approximate number of chemicals evaluated in the reported studies	Approximate percentage of correct designations	Estimate of the number of unique compounds evaluated in similar tests[a]
Ames	420	90	4000–6000
Transformation	87	90	500
Mouse lymphoma	32[b]	95	1000
SCE	47[b]	81	1000
Drosophila (SLRL)	100[c]	90	1000–2500
UD Synthesis	16[b]	87	500
In vitro cytogenetic analysis	164	75–80	1000–2500

[a] Based on estimates of published and unpublished in-house testing by private firms, including contract laboratories. It gives an estimation of the experience factor associated with the tests and an indication of their use in chemical testing. These estimates represent results on unique compounds.
[b] Number of reported chemicals insufficient for accurate determination of reliability.
[c] Polycyclic hydrocarbons are generally negative in *Drosophila*.

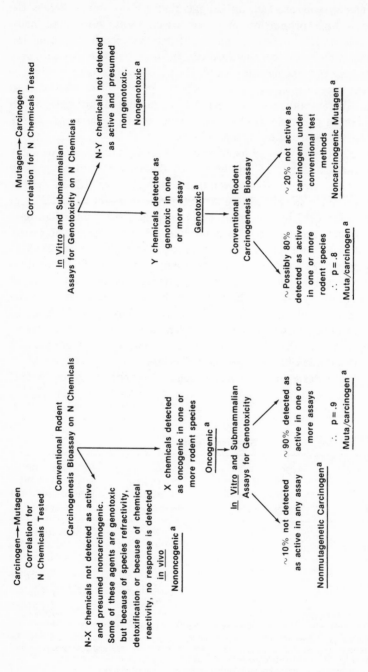

FIGURE 3.8. Explanation for the presence of noncarcinogenic mutagens (false positives). One may expect a certain proportion of chemicals, defined as mutagens by short-term tests, to be inactive when administered to the whole animal. Many of the barriers present in intact animals that reduce the effective dose of a toxicant are not present in the short-term *in vitro* assays. [a] Operational definition of agent.

studies conducted in 1962 and 1971.[21] Following a deluge of data from mutagenicity and short-term carcinogenicity testing procedures showing AF-2 to be genotoxic, it was retested in a more comprehensive rodent test and subsequently reclassified as a carcinogen in 1978. The reverse is also true. For example, DDT, which was once considered to be a rodent carcinogen, has recently been retested by the National Cancer Institute and considered noncarcinogenic. DDT is not mutagenic in most assays, especially those measuring gene mutation. Thus, terms such as false positive and false negative have questionable meaning, and can be more accurately defined as mutagenic noncarcinogens and carcinogenic nonmutagens, respectively. These descriptions should be maintained at least until adequate tests have been performed to resolve apparent conflicts.

Other, and possibly the most important factors generating mutagenic noncarcinogens are the *in vitro/in vivo* differences. If among N chemicals tested in rodents for carcinogenicity (Figure 3.8), X were identified as positive, one would have eliminated N–X compounds as either noncarcinogenic agents or agents not compatible with *in vivo* detecting systems. Agents in the N–X class would be identified as negative in tests for genotoxicity. Among the group of agents not compatible with *in vivo* testing will be chemicals that may be rapidly detoxified, are too toxic to give in sufficient doses, or are to unstable to be adequately tested in conventional rodent assays. Some of these will show positive results in simple *in vitro* assays which possess none of the *in vivo* barriers (Figure 3.9). Thus, the preselection of chemicals active *in vivo* among rodent carcinogens will produce a biased correlation in the carcinogen → mutagen direction compared to the mutagen → carcinogen direction.

The arguments against the use of genetic toxicology because of too many "false positives" are probably invalid, since the true potential of noncarcinogenic mutagens remains to be determined. For example, it would require a population of 10,000 rodents to detect a 1% increase in tumor incidence using the bioassay approach using the strains of animals typically employed. Such agents as AF-2, ethylene dibromide, or the flame retardant Tris might still be prevalent in the human environment if genetic toxicology results had been ignored because of the possibility of false positive responses.

The mirror image of false positives is "false negatives," or more accurately, nonmutagenic carcinogens. It is generally presumed that nonmutagenic carcinogens consist primarily of the agents identified as cocarcinogens and promoting agents (see Figure 3.7). These materials may act by inhibiting DNA repair of background genetic effects and/or stimulating DNA replication, thus altering the hormonal or immunologic homeostasis of the treated animal. Recent studies have shown a close correlation between the skin-tumor-promoting activity of phorbol esters and their

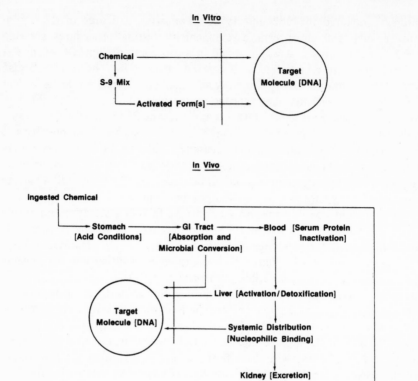

FIGURE 3.9. Factors affecting expression of genotoxic potential. Chemicals active *in vitro* will not always produce responses *in vivo* because of physiological and physical barriers.

ability to stimulate DNA synthesis.[13] In general, however, they are not electrophilic and do not react with DNA to produce lesions that may lead to mutation. There will be a few initiating agents that, because of unique *in vivo* metabolism requirements, cannot be detected as mutagens with S9 microsomal enzyme preparations.

Some carcinogen classification schemes have already been proposed which distinguish between genotoxic (DNA) carcinogens and nongenetic carcinogens. Nongenetic carcinogens are generally not active in short-term assays. This category can and has been further subdivided by Williams[25] into promoters, cocarcinogens, hormone carcinogens, solid-state carcinogens, and immunosuppressors. Agents that fall within each subgroup exhibit similar physical, chemical, and toxicologic properties. Their modes of action, however, have been hypothesized to in some manner predispose the target cells to undergo transformation or to permit the expression of already transformed cells.[7] A model for this type of tumor induction is shown in Figure 3.10.

Another feature of the two types of tumor induction is that while the experimental design to establish a true no-effect or threshold level for a mutagenic carcinogen is for all purposes impractical, a no-effect level for nongenetic carcinogens may be obtainable. The no-effect will be related to the level of exposure below that which produces physiological effects leading

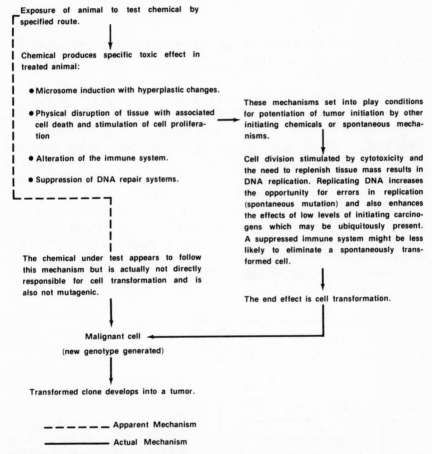

FIGURE 3.10. Proposed steps leading to tumor initiation by indirect carcinogens. Many agents, while not initiating cell transformation, can lead to the enchancement or expression of transformed cells. Some of the possible mechanisms are shown. From Brusick, D. J.: The role of short-term testing in carcinogen detection. *Chemosphere* **5**:403–417, 1978 (see also Figure 8.10).

to enhanced expression or susceptibility of the target cells. At the present time there is not a great deal of published data demonstrating the specific points identified in Figure 3.9; however, the carcinogenic mechanisms of agents such as nitrilotriacetic acid, sodium saccharin, DDT, asbestos, and other nonmutagenic carcinogens appear, in general, to also fit the scheme outlined in Figure 3.10. Testing with some of these agents, such as asbestos, results in elevated frequencies of chromosome aberrations at extremely high exposure levels.[1] This effect appears to be associated with a generalized toxicity, but may be linked in some manner to the final toxic response.

RELATIONSHIP OF POTENCY BETWEEN MUTAGENICITY ASSAYS AND *IN VIVO* ONCOGENICITY

Several investigators have begun to look at the potency range for mutagenic and carcinogenic properties of mutagenic carcinogens to determine whether they are similar or dissimilar. If similar, then *in vitro* mutagenicity studies might be useful for estimating the *in vivo* potency of suspect carcinogens. Reports by Meselson and Russell[19] and by Clive[11,12] indicate that the activity of both mutagens and carcinogens covers a 10^6-fold range and tentatively shows a roughly linear relationship between mutagen potency and carcinogenicity (Figure 3.11). Considerable debate has been stimulated on this question, and more details are presented in Chapter 5. The approach is complicated considerably by decisions on which sets of mutagenicity and carcinogenicity data will be used in the comparison. Biased selection can alter the relationship.

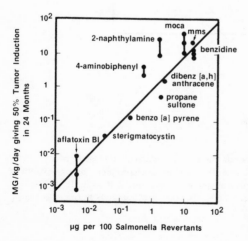

FIGURE 3.11. Relationship between potency of mutation and cancer. Several investigators have attempted to relate potency in short-term tests with potency in animal studies. While this concept is potentially very interesting, technical problems will probably prevent the general use of such comparisons. From Meselson and Russell.[19]

REFERENCES

1. Ames, B. N.: Identifying environmental chemicals causing mutations and cancer. *Science*, **204**:587–593, 1979.
2. Ames, B. N., Durston, W. E., Yamasaki, E., and Lee, F. D.: Carcinogens are mutagens: A simple test system combining liver homogenates for activation and bacteria for detection. *Proc. Natl. Acad. Sci. U.S.A.* **70**:2281, 1973.
3. Bauer, K. H.: Mutationstheorie der Geschwulst-Entstehung. Übergang von Körperzellen in Geschwulstzellen durch Gen-Änderung. Springer, Berlin, 1928.
4. Benditt, E. P.: The origins of atherosclerosis. *Sci. Am.* **235**:74, 1977.
5. Bignami, M., Morpurgo, G., Pagliani, R., Carere, A., Conti, G., and DiGiuseppe, G.: Non-disjunction and crossing-over induced by pharmaceutical drugs in *Aspergillus nidulans*. *Mutat. Res.* **26**:159–170, 1974.
6. Brewen, J. G., and Preston, R. J.: Analysis of chromosome aberrations in mammalian germ cells. In *Chemical Mutagens: Principles and Methods for Their Detection*, Vol. 5 (A. Hollaender and F. J. de Serres, ed.), Plenum Press, New York, pp. 127–150, 1978.
7. Brusick, D. J.: The role of short-term testing in carcinogen detection. *Chemosphere* **5**:403–417, 1978.
8. Burdette, W. J.: The significance of mutation in relation to the origin of tumors: A review. *Cancer Res.* **15**:201, 1955.
9. Burnet, F. M.: *Intrinsic Mutagenesis: A Genetic Approach to Aging*, Medical and Technical Publishing, Lancaster, England, 1974.
10. Cacheiro, N. L. A., Russell, L. B., and Swartout, M. S.: Translocations, the predominant cause of total sterility in sons of mice treated with mutagens. *Genetics* **75**:73–91, 1974.
11. Clive, D.: A linear relationship between tumorigenic potency *in vivo* and mutagenic potency at the heterozygous thymidine kinase (TK+/−) locus of L5178Y mouse lymphoma cells coupled with mammalian metabolism. In *Progress in Genetic Toxicology* (D. Scott, B. A. Bridges, and F. H. Sobels, eds.), Elsevier/North-Holland, Amsterdam, pp. 241–247, 1977.
12. Clive, D., Johnson, K. O., Spector, J. F. S., Batson, A. G., and Brown, M. M. M.: Validation and characterization of the L5178Y/TK+/− mouse lymphoma mutagen assay system. *Mutat. Res.* **59**:61–108, 1979.
13. Dicker P., and Rozengurt, E.: Stimulation of DNA synthesis by tumor promoter and pure mitogenic factors. *Nature (London)* **276**:723–726, 1978.
14. Hart, R. W., Setlow, R. B., and Woodhead, A. D.: Evidence that pyrimidine dimers in DNA can give rise to tumors. *Proc. Natl. Acad. Sci. U.S.A.* **74**:5574–5578, 1977.
15. Lejeune, J.: The 21 trisomy—current stage of chromosomal research. *Prog. Med. Genet.* **3**:144–177, 1964.
16. Kalter, H.: Correlation between teratogenic and mutagenic effects of chemicals in mammals. In *Chemical Mutagens: Principles and Methods for Their Detection*, Vol. 6 (A. Hollaender, ed.), Plenum Press, New York, 1977.
17. Knudsen, A. G., Jr.: Mutation and human cancer. *Adv. Cancer Res.* **17**:317–352, 1973.
18. Magee, P. N.: The relationship between mutagenesis, carcinogenesis and teratogenesis. In *Progress in Genetic Toxicology*, Vol. 2 (D. Scott, B. A. Bridges, and F. H. Sobels, eds.), Elsevier-North Holland, Amsterdam, pp. 15–27, 1977.
19. Meselson, M., and Russell, K.: Comparisons of carcinogenic and mutagenic potency. In *Origins of Human Cancer, Book C, Human Risk Assessment* (H. H. Hiatt, J. D. Watson, and J. A. Winstein, eds.), Cold Spring Harbor Laboratory, Cold Spring Harbor, N.Y., pp. 1473–1481, 1977.
20. Miller, J. A., and Miller, E. C.: A survey of molecular aspects of chemical carcinogenesis. *Lab. Invest.* **15**:217, 1966.

21. Nagao, M., Sugimura, T., and Matsushima, T.: Environmental mutagens and carcinogens. In *Annual Review of Genetics*, Vol. 12, Annual Reviews, Palo Alto, Calif., pp. 117–59, 1978.
22. Paterson, M. C.: Environmental carcinogenesis and imperfect repair of damaged DNA in *Homo sapiens*: Causal relation revealed by rare hereditary disorders. In *Carcinogens: Identification and Mechanisms of Action* (A. C. Griffin and C. R. Shaw, eds.), Raven Press, New York, pp. 251–276, 1979.
23. Setlow, R. B.: Repair-deficient human disorders and cancer. *Nature (London)* **271**: 713–717, 1978.
24. Van Valen, L.: Selection in natural populations. III. Measurement and estimation. *Evolution* **19**:514–528, 1965.
25. Williams, G. M.: The detection of chemical mutagens/carcinogens by DNA repair and mutagenesis in liver cultures. In *Chemical Mutagens: Principles and Methods for Their Detection*, Vol. 6 (F. J. de Serres and A. Hollaender, eds.), Plenum Press, New York, pp. 61–79, 1980.

CHAPTER 4

Screening Chemicals for Genotoxic Properties

INTRODUCTION AND BACKGROUND

The type of chemical evaluation needed is to a large extent dependent on the needs of the originator of the testing program. Much of the data derived from toxicology testing is used in decision-making situations for compound development. This type of evaluation then addresses the question of whether or not a test substance will be suitable for release into the environment. The suitability depends on the intended use of the material as well as the regulatory climate with which substances of this type must comply. Health and regulatory considerations must be combined with development and production costs and timing, marketing parameters, and other factors which affect the final decision.[9]

Other types of data may be needed if the decisions center on how to handle chemicals already in the environment that have genotoxic properties. In this situation, data defining possible risks following exposure will be needed. This data combined with estimates of the number of individuals exposed, the duration and level of exposure, and the cost of eliminating the material from the environment will form the basis for a different set of decisions.

Thus, it is difficult to generalize about the choice of methods for chemical testing. The two major philosophies of testing have been hierarchical (tier) approaches and matrix (battery of tests) schemes. Tier approaches have been most often applied when the number of substances that require testing is very large. The concept is that at the first level of testing, where all substances must be examined, the tests should be rapid, reliable, and inexpensive.[6,17] Materials can be eliminated on the basis of their responses at this level, and only a small number of substances will be left to evaluate in the more sophisticated, costly, and time-consuming tests. The tier concept is a cost-effective program, but it takes more time. The rationale for this approach is outlined in Figure 4.1. The first tier or phase consists of microbial tests for mutation, repairable primary DNA damage, and mitotic crossing-over phenomena. These tests permit rapid evaluation of several hundred to several thousand substances per year at relatively low cost. To be effective, screening tests should have uniform protocols, be easy to perform and evaluate, and produce relatively few positive or negative

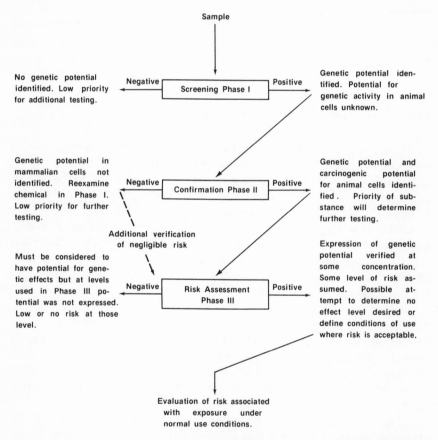

FIGURE 4.1. Tier approach to genetic and carcinogenic testing. The tier approach appears to be the most cost-effective approach to testing but often extends the testing time to unacceptable limits since each subsequent tier must wait until testing from the previous tier is completed. Specific tests to be placed in tiers are not defined and may be selected on the basis of regulatory requirements.

results that cannot be supported by tests in animal systems. The reliability of the assays used in the first level of the tier grouping is particularly critical; if these tests are not reliable indicators of genotoxic agents, the tier approach will ultimately be ineffective. Substances passed on to the second and third tiers are subjected to more testing with *in vitro* mammalian cell assays, *Drosophila*, or *in vivo* mammalian models. The intent of the evaluation at the highest level (Tier III) is to develop an assessment of possible human risk from anticipated exposures.

Increasingly, the direction of testing programs is shifting from a multi-phase tier approach to a test battery approach. There are several reasons for the shift. Timing of data collection, concern over false positive or negative responses from Tier I screening tests, and the need to measure multiple end points all have contributed to the battery philosophy, as have the recommended mutagenicity guidelines from the Environmental Protection Agency (EPA) and other government agencies. A battery of tests includes several levels of evaluation conducted simultaneously to provide a window through which nongenotoxic agents can pass, but which allows chemicals likely to produce mutations or cancer in animals to be identified. Figure 4.2 illustrates the battery approach to testing. Each test in the battery has its own range of detection when used with a wide range of chemicals. A group of tests should overlap in response for most genotoxic agents, but not all; thus, unanimous agreement across all tests included in a test battery should not be expected. On the other hand, each test may detect a small proportion of effects not predictive of *in vivo* responses. Because of the technical limitations of the current state of the art, some genotoxic agents will be missed by

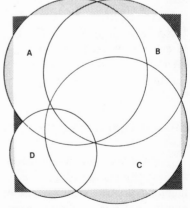

FIGURE 4.2. Battery concept. Each short-term test has its own inherent strengths and limitations which, when used in concert, will overlap and compensate for intrinsic limitations. The square represents the total number of chemicals with mutagenic and/or carcinogenic activity. Each circle represents a specific test and its range of detectability. Overlapping indicates compounds detected in multiple assays. The area of the circles will vary according to the test selected.

▨ Responses designated as false negatives

▨ Responses designated as false positives

even the best test battery. However, careful selection of tests will keep this to a minimum.

Test batteries may consist of screening tests, risk assessment tests, or a mixture of both. It is important at the outset of testing, however, to carefully define the biological question being asked. Screening tests are those that measure the ability of the test substance to produce genotoxic effects. Results from screening tests alone cannot be used in risk estimation, but can define the genotoxic potential of an agent. No attempt is made to relate dose, route of administration, or metabolism to the human experience. Risk assessments require *in vivo* data that may be synthesized with *in vitro* data. Claims of potential health hazards are difficult to justify if based solely on *in vitro* or submammalian test results. This is not to say that it has not been done in the past, but proponents of this application of screening tests have created confusion and in some instances resistance to the acceptance of genetic toxicology.

There may be justification, however, for the development of hazard assessment using results from combined submammalian, *in vitro* mammalian, and *in vivo* somatic tests.

1. The current state of the art in genetic toxicology is not sophisticated enough to accurately assess point mutation induction in mammals, and especially in mammalian germ cells. The most practical of the *in vivo* models rely upon measuring chromosome alterations.[35] The combined use of submammalian and *in vitro* assays is necessary, therefore, to build a comprehensive data base for the genotoxic properties of a particular test sample.

2. The target molecule of all genotoxic effects, DNA, is structurally and functionally the same in all organisms. If the test sample is capable of altering the structure or function of DNA in a submammalian species, it can also be expected to produce the same response in rodent and human DNA, providing the biologically active form of the test sample reaches the target site in each case. Thus, *in vitro* mutagenicity data coupled with *in vivo* chemical disposition data showing germ cell interactions may, under certain circumstances, be a viable approach to an estimate of genetic risk.

CHARACTERISTICS OF ADEQUATE SCREENING TESTS

Since the battery-approach concept appears most consistent with the natural application of genetic toxicology testing, the characteristics of tests included in the various types of test batteries need to be examined. The number of tests available for inclusion in a battery is extensive and covers many divergent phylogenetic levels. Not all of the tests available are considered screening tests. Most of the mammalian *in vivo* tests, for instance, with the exception of some host-mediated types and possibly the micronucleus test, have application to risk estimation for somatic and germ cell effects. Screening implies a preliminary assessment prior to critical evaluation, and this appears to be the primary role for the bulk of the submammalian and *in vitro* tests.

As indicated in the following list, there are several features of a bioassay that make it a candidate as a screening test.

1. The test should specifically identify agents with affinity for DNA, and not agents with nonspecific toxicity for macromolecules. DNA interactions can be measured as primary DNA effects, point (gene) mutations, chromosome alterations, and possibly morphologic transformation.

2. The test should have adequate inherent capacity for metabolic biotransformation, or should be capable of incorporating one of the standard activating systems described in Table 4.1.

3. The test should be sufficiently validated so that interpretations of both positive and negative data have meaning in estimating the hazard or safety of a substance.

4. Adequate facilities to conduct the test should be available in several independent laboratories. Tests capable of being performed only in a single laboratory are probably not adequate for routine screening.

These features are discussed in detail on the following pages.

The Type(s) and Number(s) of End Points Detected

Chemicals alter DNA in both specific and nonspecific ways. Point mutations and chromosome rearrangements are rather specific classes of genotoxic end points. Chemicals also modify DNA in such a way that neither point mutations nor chromosome aberrations result. This type of nonspecific genotoxicity has been referred to as primary DNA damage. It usually results in cell death unless the prelethal lesions are repaired. Two methods of detecting primary DNA damage are by stimulation of the repair processes or by loss of prelethal damage.

Another end point not generally classified under DNA-specific damage but which appears to result from multiple steps is *in vitro* cell transformation. The end product of cell transformation is a cell mass (focus or clone) with altered morphology, growth patterns, and biochemistry. Transformed cells are also capable of proliferating into a tumor when implanted into the appropriate host animal.[30,33] The transformation assay is considered a close approximation of the events which occur *in vivo* during oncogenesis.

Thus, it is advantageous to know which type of genotoxic end point an assay measures, and whether a specific assay is capable of measuring more than one end point. Systems measuring more than one end point are highly preferable for inclusion in a test battery. It is also important that the tests

TABLE 4.1

Types of Chemical Activating Systems That Are Used with Screening Tests

Activating system	Reference	Advantages/limitations of the method

Host-mediated assays (HMA)
  a. Intraperitoneal
  b. Intrasanguine
  c. Urine analysis
  d. Other
	
	18
	4
	36
	25
		All have the advantage of reflecting the typical *in vivo* activation/detoxification balance. Their major problem is associated with location and recovery of indicator cell within the host.

Microsomal enzyme preparations
  a. Purified microsomes
  b. S20 preparations
  c. S9 preparations
  d. Crude homogenates
	
	20
	26
	1
	[a]
		The primary advantage lies with the fact that good contact between chemical and enzymes can be assured in the microsome mix. The response interpretations are complicated by the loss of all compound activation/detoxification interactions as well as the coordination in conjugating and deconjugating enzyme systems.

Cell-mediated systems
  a. Primary hepatocytes
  b. Irradiated feeder
     cells
	
	38
	22
		These systems provide good metabolic balance and preserve the intact metabolic systems as found in the intact liver. They also maintain close proximity between test substance and activating enzymes. The cell-mediated systems do not require cofactor reaction mixtures. The limitations rest with the technical difficulties of preparing the cell systems and coupling them to the test methodology. This approach appears most promising with mammalian cell culture techniques.

[a] E. Zeiger, unpublished data.

selected for a battery do not all measure the same end point. All genotoxicity is potentially hazardous, but as a general rule the hierarchy of end points places the significance of point mutation and transformation greater than chromosome effects, which in turn are greater than primary DNA damage. The basis of this ranking is the correlation of the end point to *in vivo* point mutation or oncogenesis. There are difficulties associated with interpreting chromosomal and nonspecific damage that are likely to result only in lethality.

The Metabolic Capability of the Test or Associated Activating System

The application of the HMA and the S9 microsome systems to screening has enhanced the detecting capacity of most tests. Some test, such as

Drosophila and some of the fungi, have intrinsic metabolic systems.[10,42] Most microbial and *in vitro* mammalian cell assays, however, must be coupled with some type of activating system before they can be used for screening (Table 4.1).

The subject of *in vitro* metabolism is complex and controversial. *In vitro* microsome-mediated metabolism offers an isolated view of the biological properties of one or more activated forms of the test substance but often fails to put the products into perspective relative to *in vivo* metabolic pathways. This subject is dealt with in many review articles and books and will not be discussed in detail here.[14,27] However, a great deal of work has been conducted with microsomal activation systems, and it is safe to say that they provide a rough approximation of *in vivo* events, although this approximation is not so accurate that reliable quantitative extrapolations can be made from tests employing the systems. The use of isolated fresh hepatocytes probably gives a better representation of *in vivo* effects than microsome fractions, especially when detoxification processes are of interest. In S9 mixes, the balance between activation and detoxification favors activation, since several detoxifying enzymes such as glutathione transferase are inoperative.

The following generalizations regarding the various activation systems are probably accurate.

1. HMA systems are more likely than *in vitro* microsome preparations to generate false negative results.

2. Interspecies comparisons of microsomal preparations have shown clear quantitative differences in metabolic activation, but very little evidence exists for unequivocal qualitative differences. This includes studies with human S9 preparations which indicate that rodent S9 preparations are reasonably good models for human activation reactions.[40] However, most of the species differences disappear when hepatic inducers such as PCBs are used in S9 preparation (Appendix A).

3. Although various methods of microsomal enzyme induction have been reported, the use of Aroclor 1254 produces a broad-spectrum induction and seems to be adequate for detecting most genotoxic substances. Induced microsomes are preferred over noninduced preparations for routine testing.

4. There does not appear to be any advantage in using microsome preparations cleaner than S9 (e.g., S20 or purified microsomes), and crude tissue homogenates appear less optimal than S9 preparations.

5. Little benefit has been derived from the use of activating enzymes from tissues other than liver. While quantitative differences can be observed qualitatively between tissues, the results with liver cover activity detected in other organs. The same conclusion appears valid across mammalian species. Again, the variation in activation appears more quantitative than qualitative. These considerations are true when related to screening chemicals;

however, quantitative differences become important in studies focused on risk. It may be desirable, in some instances, to match species between *in vitro* activation and toxicity data collected *in vivo* if mutagenic metabolites are encountered.

Reliability/Reproducibility

There has been considerable concern among scientists over the potential problems associated with intra- and interlaboratory reproducibility of test results. This is especially true for the submammalian and *in vitro* mammalian assays. There are a number of reasons for this concern, some legitimate and some not. It is possible for two laboratories to disagree on the genotoxic potential of a given substance, or in some cases for the same substance to be declared both active and not active on two different occasions using the same test in the same laboratory. The scientific literature can be used to document inconsistencies for all types of toxicologic testing. Significant intra- and interlaboratory variability has been encountered in laboratory findings ranging from acute LD_{50} (lethal dose 50%) measurements to chronic carcinogenesis bioassays in rodents. The real question is does it happen more frequently in submammalian and *in vitro* tests for genotoxicity? The answer appears to be that it does not. One reason for the belief that tests routinely used in genetic toxicology are more prone to nonreproducibility is that more substances are being subjected to *in vitro* testing, often in multiple laboratories. Thus, it is reasonable that inconsistencies apper to be more common. Compared with the overall intra- and interlaboratory consistency, the differences are relatively infrequent.

If the potential problems of *in vitro* and submammalian tests were listed, one item might be the almost unlimited access to the test system by anyone wishing to conduct an evaluation, especially individuals with little or no training in genetics. Such is not the case for large-scale, animal-based toxicology studies. For example, the *Salmonella* strains used in the Ames assay are in virtually all microbiology laboratories in the United States. The cost of conducting the test is minimal, and the temptation to put all types of substances through the test is strong. Superficially, this may not represent a problem, but anomalies will certainly arise without standardized test methods, quality-control procedures, and data-recording and data-evaluation criteria. If the individuals conducting the test are not thoroughly experienced with the assay, it is probable that such anomalies will not be identified, and the aberrant results will be reported. Once results are reported in the scientific literature or to regulatory agencies, it is extremely difficult to remove their impact, even if it can later be established that the effects were the result of testing artifacts. Some control or verification efforts must be put into practice to prevent the unconscious misuse or misinterpretation of short-term tests and their results.

The enactment of Good Laboratory Practices Regulations by the U.S. Food and Drug Administration (FDA) (see Chapter 7), as well as some recently proposed certification and accreditation programs for testing laboratories and personnel, may be the mechanism. These efforts should produce a tremendous impact on data consistency. They will not eliminate anomalies in testing, but should improve the overall performance, standardization, and review of data.

Another important factor in the production of apparent conflicting results is the method of data evaluation and interpretation. Similar sets of data may be evaluated differently by two investigators, and if the data are marginal they might be considered indicative of an effect by one and no effect by the other. Each investigator's view of the data will be affected by past experience with the test system and the type of evaluation criteria and statistical analysis used to interpret the data.

The size and, therefore, the power of the test also contributes to inconsistent interpretations. Comparison of data from two studies, one using minimal sample sizes and one using a large-scale study design having a greater resolving power, may yield different conclusions. Minor variations in scored events in a small-scale study may produce statistically significant data points which would be shown to be irrelevant in a more powerful test. It is therefore extremely important in making intra- and interlaboratory comparisons that careful analysis of the data is accomplished by thoroughly examining the comparability of study designs, the sample size, and the scoring procedures employed.

Even when all factors of studies are controlled, there will be occasional sets of conflicting test results. For this reason it is recommended, when feasible, to incorporate an automatic, confirmatory retest sequence into chemical screening programs. If all or a portion of each test, regardless of the response, is repeated after a specified interval following the first test and the results match the interpretation, the possibility of test anomalies or experimental error accounting for the results is significantly reduced. If the results are conflicting, additional testing will be necessary to resolve the conflict. Obviously, this is a more costly and time-consuming approach to screening, but the benefits derived from reproducible results will increase the reliability of the data base and facilitate scientific and/or regulatory decisions. This approach will also be beneficial in that a reproducibility or reliability quotient can be ascertained both for the test and the laboratory. As tests gain more extensive data bases, these reliability quotients will become important in selecting the most suitable test methods. Some components of reliability quotients are:

1. The test should be highly reproducible both within and among laboratories.
2. It must be documented that a test measures true genetic lesions at the DNA level.
3. Methods of scoring and analyzing data need to be defined and justified.
4. The level of spontaneous or background events must be well defined and included in the definition of an adequate test.
5. Scoring methods should be as objective as possible and not subject to unintentional bias.
6. The experience of the test with a broad range of chemical classes should be sufficient to give meaning to a negative finding.
7. The weaknesses and limitations of the test system should be known and factored into the data interpretation.
8. There should be specific criteria to define an adequate test.

Currently, the reliability of only a relatively few assays can be adequately documented. These consist of the Ames *Salmonella*/microsome assay, the SLRL assay in *Drosophila*, rodent bone marrow cytogenetic analysis, the rodent dominant lethal assay, and the L5178Y mouse lymphoma TK forward mutation assay. There has been a formal interlaboratory comparative study for each of these tests. Reasonably good agreement has been obtained among the assays mentioned, with the only exception being the rodent bone marrow cytogenetic analysis. Collaborative cytogenetic studies suggest that chromosome scoring techniques are usually more laboratory-dependent or investigator-dependent than other testing methodologies (see Figure 6.1). Chromosome scoring, like histopathology scoring, is somewhat subjective; and individual differences in cytologic interpretation are likely to occur. It appears that most if not all of the screening tests that involve cytologic examinations as an integral part of the scoring procedure exhibit the highest levels of interlaboratory variability.

The exception to this subjectivity appears to be the SCE assay. SCE scoring is highly reproducible and nonsubjective compared to aberration analysis, and ranks very high on the reliability scale in limited comparisons. The interpretation of the toxicologic significance of SCE has yet to be resolved, and thus this test, even though less subjective, cannot replace conventional cytogenetic analysis.

Facilities

Facilities involved in genetic toxicology should be adequately staffed by experimental investigators and technicians, and should be in compliance with Good Laboratory Practices and accreditation standards. Further discussion of genetic toxicology facilities can be found in Chapter 7.

An assay that is conducted in only a single facility is probably not adequate for routine screening, since there would be no opportunity for independent confirmation or review of test results.

STRATEGIES FOR TEST BATTERY DEVELOPMENT

General Philosophy

There can be a variety of themes on which to base test selection. For the most part, each battery represents the experience and bias of the investigator. Some batteries are built around carcinogenicity prediction; others are broader and try to encompass rodent mutagenesis. A recent comparison using carcinogenicity as the standard has been compiled (V. Ray, personal communication). The comparison was based primarily on a survey of published data for 511 chemicals in a series of frequently used genetic toxicology tests. Several conclusions were reached from this comparison.

1. Most, if not all, chemical carcinogens having genotoxic properties would have been identified by a battery consisting of the Ames *Salmonella*/microsome assay, the mouse lymphoma assay, cell transformation in the SHE assay, and *in vitro* cytogenetics.

2. Most *in vivo* tests such as the bone marrow cytogenetics and micronucleus tests, the rodent dominant lethal test, the rodent heritable translocation test, and conventional HMA are poor predictors of mammalian carcinogenicity.

3. The actual testing experience with many candidate tests is extremely limited and in some cases restricted to tests performed in only one or two laboratories.

4. Much of the published data is not sufficiently documented to draw reliable conclusions. Conflicting data cannot always be resolved because of inadequate descriptions of experimental designs or presentation of raw data.

Approaches to Test Battery Development

In addition to the literature review and comparisons developed by Ray, other programs of test comparisons are in progress. The largest of these programs is the Gene–Tox program sponsored by the EPA. The purpose of the Gene–Tox program is to generate critical reviews of the test systems listed in Table 4.2. These reviews are to be written by groups of experienced scientists who will summarize and prepare reports on the published data base for each usage. The input from the working groups will be used to (1) evaluate the utility of each test based on the present data base, (2) identify which tests perform best under present methodologies on the broadest range

TABLE 4.2

Assays under Review in the Gene–Tox Program Sponsored by the U.S. Environmental
Protection Agency

Effect	Test system
Gene mutation	*S. typhimurium*
	E. coli WP$_2$
	Mouse lymphoma cells L5178Y
	Chinese hamster lung cells V79
	CHO cells
	D. melanogaster
	Neurospora crassa
	Aspergillus nidulans
	Tradescantia
	Mouse specific locus (including spot test)
	HM Assay
Chromosomal effects	*D. melanogaster*
	Plant cytogenetics
	Saccharomyces cerevisiae
	Schizosaccharomyces pombe
	Mammalian cytogenetics
	SCE
	Micronucleus test
	Dominant lethal test
	Heritable translocation test
Primary DNA damage	*E. coli* polA^+/A^-
	B. subtillis "rec" assay
	UDS
	DNA repair in mammalian cells
Oncogenic transformation	C3H/10T½ mouse fibroblasts
	BALB/c 3T3 mouse fibroblasts
	SHE
Ancillary test	Mouse sperm morphology

of chemicals, (3) develop protocols for testing, (4) identify tests which need more validation efforts before they can be applied to routine testing, and (5) select test batteries most suitable for given classes of chemical mutagens and carcinogens. The results of the Gene–Tox program, while not currently available, will potentially have a major impact on the manner in which chemicals are screened. The International Commission for Protection against Environmental Mutagens and Carcinogens (ICPEMC), founded in 1977,* is also taking an active role in reviewing the applicability of test methods on an international level.

---

* For details regarding the goals and composition of this commission, see *Mutat. Res.* **54:**379–381, 1978.

Another recent review of the performance of short-term tests as predictors of animal carcinogenesis has been published by Hollstein et al.[21] The results of this review, which involves only 72 substances, essentially confirm the comparison developed by Ray.

Federal regulations or guidelines (FIFRA, TSCA, etc.) for genetic toxicology advocate a multitest approach consisting of in vitro and in vivo assays. Chemicals already present at a significant volume in the environment would probably be subject to an extensive test program, including screening and risk assessment assays. However, the cost of full-scale animal testing for mutagenic and carcinogenic properties may be an economic deterrent for new substances being developed. Thus, a two-step approach using screening tests to give a preliminary assessment of toxicity followed by in vivo studies on priority compounds might provide satisfactory safeguards against detrimental health effects while retaining cost-effective programs for evaluating new chemicals. The basis of this approach is a simplified tier approach using two primary test batteries plus a third, or supplementary, group (Table 4.3).

The first battery of tests (Group I) should consist of submammalian and mammalian in vitro tests that have proven capabilities for detecting chemicals with high probabilities of demonstrating mutagenic and carcinogenic activities in mammals. This battery of tests should be sensitive, rapid, and relatively economical and should contain types of assays which can be conducted in most currently existing testing laboratories. This battery should detect all substances which can produce DNA toxicity of the types identified in Chapter 2.

The utility of Group I tests is tied to their predictive capacity for mammalian mutagens and carcinogens. In a sense they serve two functions. While numerous candidate assays could be listed for each genetic type in

TABLE 4.3
Simplified Tier Approach

Group I—Screening for genotoxic agents

This type of test defines the potential for a substance to produce genotoxicity. Submammalian and mammalian in vitro assays are routinely employed.

Group II—Estimating risk from exposure to genotoxic agents

This type of test measures the expression of genotoxicity in vivo under conditions similar to the anticipated human exposure. Mammalian in vivo model systems are employed.

Group III—Supplementary tests

These tests are used in special situations. Their applications are supplemental to the tests in Groups I and II.

Group I, actual performance under routine testing conditions has shown that certain tests will fill the need better than others.

The second battery of tests (Group II) will be used on priority compounds selected from those evaluated in Group I tests. The Group II tests should be able to estimate the likelihood that genotoxic activity detected in Group I tests will be expressed in animals under anticipated exposure conditions. Group II tests will require considerably more time and expense. Their primary role would be to estimate genetic risk.

Tests that supplement Group I and Group II, but may be somewhat too specialized or complex for general application, are grouped into a supplementary battery (Group III). These tests could be used as additions to either Groups I or II in specific cases, or to meet regulatory guidelines requiring testing redundancy.

It can be argued from *in vivo* data bases that substances adequately evaluated by Group I tests and found to be uniformly negative are very unlikely to demonstrate either mutagenic effects in Group II tests or carcinogenicity in rodent bioassays, and little will be gained by additional testing. Substances not active in Group I tests should have a very low priority for further evaluation. Substances producing positive effects should be further examined using *in vivo* studies under some type of logical progression. Results of both sets of tests could then be applied to associated toxicologic data for the estimation of human risk. Tables 4.4 and 4.5 outline

TABLE 4.4
Group I Tests: Selection and Justification

Types of tests
  Microbial test for gene mutation
  Mammalian *in vitro* gene mutation test
  Mammalian *in vitro* assay for morphologic transformation
  Mammalian *in vitro* cytogenetic test (aberrations and SCE)

Advantages
  Cost- and time-effective
  Available laboratory resources
  Reasonably uniform protocols and evaluation criteria
  Good predictors for mammalian carcinogens and mutagens
  Measure broad spectrum of genotoxic end points
  Can be conducted with a metabolic activation system

Estimated cost per sample evaluated
  $12,000–18,000 without confirmation retest

Estimated time required for evaluation
  10–12 weeks without confirmation retest

Estimated current U.S. capacity
  Sufficient private in-house and contract laboratories to evaluate approximately 3500–4000 samples per year (without confirmation)

TABLE 4.5
Group II Test: Selection and Justification

Types of tests
  In vivo bone marrow cytogenetic analysis
  Analysis of excreted urine for genotoxic agents
  Dominant lethal assay
  Recessive somatic mutation assay for coat-color spots*
  Heritable translocation assay (HTA) in mice*

Advantages
  May be able to develop an estimate of risk
  Can replicate human exposure conditions and route of administration
  Amenable to statistical analysis
  Somatic and germ cell effects measured
  Chromosome effects and presumed point mutations measured
  Reasonable data base (except those marked with *)

Estimated cost per sample evaluated
  $35,000–45,000 excluding HTA

Estimated time required for evaluation
  5–6 months (excluding the heritable translocation assay)

Estimated current U.S. capacity
  Sufficient private and contract laboratories to evaluate 200 samples per year (excluding those marked with *)

the tests and associated requirements of time and cost for Groups I and II, respectively, and Table 4.6 outlines the tests for Group III.

Only the generic types of tests assigned to these groups have been identified, and while this composition may not represent batteries that would receive unanimous scientific approval, appropriate adjustments in test composition could be made without compromising the overall evaluation process. In the final analysis, the selection of specific tests will be based on their performance in actual testing situations, and the ability of a negative finding in each test to ensure a lack of genotoxicity. This approach to battery structure will require a thorough examination of published literature and careful documentation of the results reported. Some of the tests identified in Groups I, II, and III have limited data bases, but are included because they have the characteristics desired for this type of testing. The use of this approach should give great flexibility in resource application and in the overall management of safety testing of the high number of chemicals that will have to be evaluated.

Recommendations for Group I Tests

The goal of screening is to detect samples with genotoxic activity. This activity may be evident as primary DNA damage, point mutation, or chro-

TABLE 4.6
Group III Tests: Selection and Justification

Types of tests
  Plant mutation assays
  D. melanogaster—various end points
  UDS in mammalian cells
  Yeast cells—various end points
  Micronucleus assay
  Bacteria differential DNA toxicity
  Mouse spermhead abnormalities
  HMA

Advantages/limitations
  Can provide additional confirmation for Group I tests
  Some end points specialized and not easily interpreted
  Some tests limited in availability
  Technical limitations evident in some tests
  Data base limited for some tests

Cost and availability
  Costs of several of these tests are not well defined. The range for tests on bacteria to those
  on *Drosophila* (submammalian) is $200–15,000. The whole-animal tests and *in vitro* cell
  culture assays range from $3000 to $5000. Availability varies considerably on most of
  these assays.

mosomal effects. The selection of tests, therefore, should cover these end points. This cannot be accomplished with a battery of microbial tests because of the inability to detect chromosomal effects in these organisms. Consequently, Group I tests must contain nonmicrobial tests. The distribution across various phylogenetic levels is important to eliminate the possibility of organism-specific responses.

Another factor involved in test system selection is that of redundancy. Should a single test cover each end point, or should more than one test be employed? This, to a certain extent, depends on the substance being tested; but, in general, an adequate battery of tests should contain some redundancy. Redundancy tends to confirm or cross-check responses among tests measuring similar classes of genotoxic events, and there are numerous examples of agents missed in one test for gene mutation but detected in others, as shown in Table 4.7. The examples shown in the table support the battery concept as illustrated in Figure 4.2, and are usually cited as justification for redundancy. Chromosomal methods are not considered to be as organism specific as gene mutation at selected loci, and thus one cytogenetic test is usually sufficient for screening purposes. From a practical as well as scientific point of view, cultured mammalian cell lines or human peripheral blood lymphocyte cultures seem most suited to routine analysis for alterations; rodent or human cells are most typically employed. The advantage of the rodent systems over the human lymphocyte test is that an *in vitro* microsome activation component can be added to these cell lines. Lymphocytes are not amenable to S9 use. This problem can be alleviated by employing cultured cell lines derived from human tissue such as WI-38 cells or other human fibroblast cultures.[13] Most of these lines are compatible with the S9 microsomal enzyme activation systems. The data from cytogenetic investigations do not suggest any significant degree of difference between species or cell lines with respect to response to a genotoxic or clastogenic agent. There are, however, some agents which appear to be specific for chromosome alterations rather than point mutation induction, and it is for this reason that some type of chromosome test should be included in a screening battery. Spindle poisons and agents which alter pH and osmotic conditions in cells (e.g., colchicine, benzene, benzimidazoles, urethane, and maleic hydrazide) are most easily detected with cytogenetic assays.

A test to detect SCE induction is often recommended for screening programs. SCE effects are visualized at the chromosomal level but are not completely analogous to aberrations and not necessarily induced by clastogens. In fact, data reported by Carrano *et al.*[11] indicate that SCE tests are more likely to be redundant for point mutation tests than for chromosome tests. Thus, one might include an SCE test in a battery for broader coverage of several types of DNA lesions. In fact, SCE tests are often categorized under the general end point heading of primary DNA damage.

Other tests classified as detecting primary DNA damage (UDS, alkaline elution, differential toxicity, and mitotic recombination) were not included in the Group I recommendations primarily because of technical problems which often arise relative to data analysis and interpretation (see Chapter 8). They are placed in Group III as ancillary tests. These tests may

TABLE 4.7

Examples of Some Environmental Agents Likely to Be Missed in Group I Assays without Redundancy

Compound	Missed in	Detected in	Type of end point induced
Hexamethylphosphoramide	Ames	Mouse lymphoma	Gene mutation
Natulan	Ames	Mouse lymphoma	Gene mutation
Acrylonitrile	Ames[a]	*E. coli* WP$_2$	Gene mutation
Diethylstilbesterol	Ames	Mouse lymphoma	Gene mutation
Urethane	Ames and Mouse lymphoma	*Drosophila* SLRL	Gene mutation
Formaldehyde	Ames[a]	Mouse lymphoma	Gene mutation

[a] Some protocol modifications have been reported to increase the utility of this assay for specific compounds.

be valuable for specific types of chemicals, but in general are recommended as ancillary tests to be conducted in addition to the suggested Group I or II tests.

The third type of test recommended for Group I does not measure a genetic end point directly. Cell transformation, as it is operationally defined for *in vitro* techniques, is a change in the morphologic and growth characteristics of the target cells. Although genotoxic events are probably essential to produce transformed cells, the quantitation of transformation assays is not clearly compatible with the single-hit dose–response curve typical of mutation induction. The process is likely to involve multiple steps (see Chapter 8).

Analogous to *in vitro* cytogenetic analysis, a number of different target cells are used in transformation assays. Most common are rodent fibroblast cell lines such as the BHK-21, BALB/c 3T3, C3H, 10T½, and high-passage Fischer rat embryo cells; these cell lines have many of the characteristics of normal rodent cells. They have typical morphology, are contact inhibited, will not grow to produce a tumor when implanted into a compatible host animal, and either grow poorly or not at all in soft agar. Transformed cells, on the other hand, lack these attributes and behave differently.

The transformation system introduced by Berwald and Sachs,[5] developed by DiPaolo *et al.*,[16] and more recently refined by Pienta[32] employs primary or low-passage embryo cells from the Syrian hamster. This assay is considered by many to be the most relevant transformation model, but some technical and scoring difficulties prevent it from being used by more laboratories.

All of the transformation systems are difficult to maintain and require constant attention to function well. For that reason, selection of the transformation test to be included in a test battery must be undertaken with caution. Studies with BALB/c 3T3 cells can be conducted with relative ease, and the cells are not difficult to handle. In most cases, data from the BALB/c 3T3 test have coincided with data from the hamster embryo system.

It has long been argued that *in vitro* screening techniques use fibroblast cells, and this cell type may not be representative of the epithelial cell type from which most tumors develop. Success in developing methods employing epithelial cells has been minimal, primarily because of the absence of good criteria for identification of transformation. Use of human cells in transformation assays has also been limited. The ability to produce transformed human cells has been an elusive goal, although some investigators recently have reported successful transformation of human cells.[24,29] This is the type of test where use of human cells as the target would be beneficial in making extrapolations, since there appear to be some minor qualitative differences between species for induction of transformation (Virginia Dunkel, personal communication).

It is not likely that any single uniform cell line or type will be selected for Group I mammalian *in vitro* tests. Numerous choices are available; selection should depend on the type of end point measured, the published data base for the cell line, the previous experience of the investigator, and facilities available.

INTERPRETATION OF DATA FROM SCREENING TESTS

The selection of protocols or study designs is equally as important as the proper selection of tests for the screening program. Study designs must meet the requirements defined by Good Laboratory Practices requirements, and should be of sufficient quality to give good experimental resolution. Most submammalian and *in vitro* tests can provide good resolution because of the large number of target cells exposed to the test substance. A series of reasonably standard study designs fulfilling Good Laboratory Practices requirements is provided for reference in Chapter 9.

Special emphasis should be given to preliminary toxicity testing and dose (concentration) selection. Improper test concentrations appear to be the most consistent deficiency in genetic toxicology testing. Care must be taken to ensure that test concentrations are high enough to demonstrate physiologic changes indicative of target cell exposure, yet not so high as to produce nonspecific cellular toxicity uncharacteristic of true genotoxic agents. *In vitro* mammalian cell cultures, for example, are particularly susceptible to chromosomal disruption caused by nonspecific effects at high concentrations of agents which alter the osmotic or pH parameters of the culture medium. It is not possible to fix the dose range for each assay, but Figure 4.3 illustrates the area on the toxicity curve from which test concentrations for *in vitro* assays might be selected. The number of points selected to cover this range will depend on the nature of the test and the desired resolution. Generally, the more points sampled, the more reliable the results. Some genotoxic agents may have a narrow range of activity which coincides with solubility, pH, or other factors.

Figure 4.4 shows some examples of hypothetical dose response curves, and identifies some of the problems encountered with the hypothetical concentration series shown. For example, the type of response generated in (A) might easily be missed if the doses are spread too widely. In (C), the actual dose–effect portion of the curve is missed even at the lowest concentrations, making interpretation difficult since all test samples were taken from the plateau. Additional interpretation problems can be en-

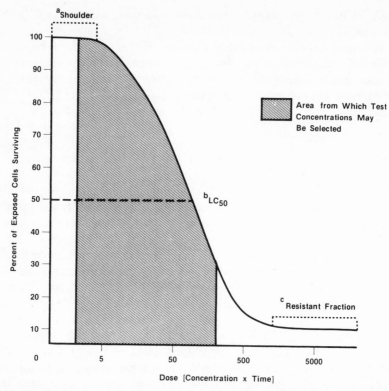

FIGURE 4.3. Typical survival curve for target cells used in screening type assays. Preliminary toxicity studies generate survival curves and are useful in selecting test concentrations. [a] Shoulder probably represents area of effective repair of all lesions. At least one concentration should fall into this region of nontoxicity. [b] Lethal concentration for approximately 50% of the target cells. [c] The resistant fraction may represent a unique subpopulation and the concentration range should not include this region.

countered if there are several doses located on the down-side portion of the curves in (A) and (B). These responses appear as an inverse dose relationship. The illustrations identify how important it is to carefully select concentrations toward the lower end of the maximum tolerated concentration level. Recommendations for selection of *in vivo* dose levels are given in Appendix B.

Since *in vitro* tests will be conducted under conditions of maximum tolerated concentration, one can expect to encounter situations in point mutation assays where the induced mutation frequency is a function of decreasing survival and not the production of an absolute increase in mutants. In this situation the possibility of preferential selection exists; that is, the test material is preferentially killing the nonmutant cells, resulting in an artifact which resembles mutation induction. An approach to determine if selection is occurring is to perform lethality tests on separate populations of normal and mutant cells with the test substance. If selection is occurring, the normal cells should show enhanced sensitivity and an LC_{50} at a lower concentration than the mutant cells. There are very few documented cases of preferential selection. The illustration in Figure 4.5 shows what may hap-

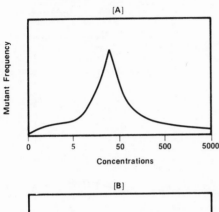

Very narrow range which may be missed if concentration steps are too far apart. Not very typical but can occur if pH range or osmotic conditions are critical. The compound is basically nontoxic.

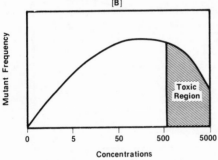

Most typical type of response curve. One may get a set of data with an inverse concentration response if dose range is selected too high.

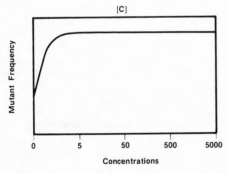

Some chemicals have only a short range over which a dose response is observed. A plateau is reached because of solubility, cell permentivity restrictions or other rate limiting processes. The compound is basically nontoxic.

FIGURE 4.4. Types of dose–response curves obtained in screening tests. Three relatively typical types of dose–response curves are shown in this diagram. The variations in shape emphasizes the need to employ multiple dose protocols for chemical screening.

Concentration [µg/ml]	Surviving Cell Fraction [x 10^4]	Actual Count of Mutants Recovered	Calculated Frequency [x 10^{-6}]
0	312	23	7.37
1	287	21	7.31
3	190	20	10.52
10	113	19	16.81
33	58	17	29.31
100	40	18	45.00
333	26	13	50.00
1,000	7	3	42.85

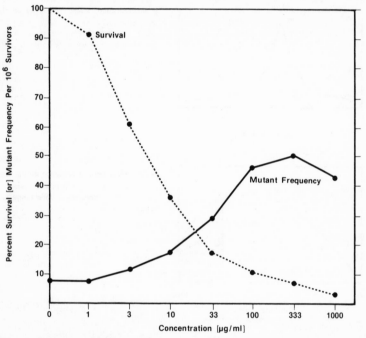

FIGURE 4.5. Potential selection error in mutation dose–response study using a hypothetical situation. It is possible to produce dose–response mutation induction curves through improper dose selection. The mutant frequency shown appears to increase with dose, but close inspection of the test data shows that there was no net increase in the number of mutants. The frequencies increased because of an apparent faster rate of killing for the nonmutant cells. This data raises the possibility of selection rather than mutation. Careful evaluation of data is required in cases of this nature.

pen if studies are conducted under conditions of extreme toxicity and low cell survival.

The Use of Controls in *in Vitro* and *in Vivo* Testing

Most, but not all, toxicology studies include concurrent controls. The most typical controls are those for the solvent or vehicle used in the preparation and administration of the test substance. In addition to the solvent/vehicle control, other negative controls (assay system minus solvent/vehicle and test substance) are occasionally conducted to determine the true background level of the assay system. Depending on the type of assay and the inherent variability, it is also possible to compare the test results with a composite historical control data base. This comparison can be valuable for certain types of studies, especially when small sample sizes are involved. It is not uncommon to also conduct a positive control. A positive control consists of performing a study similar to that used for the test sample with an agent known to produce an effect in the assay system. There are several purposes for conducting positive controls. Table 4.8 summarizes the most commonly conducted control tests.

Negative and solvent/vehicle controls have obvious importance in conducting toxicology tests. These data provide the no-effect or baseline response from which all statistical or other types of data interpretation emanate. Reliable data analysis would be impossible without the inclusion of these controls in a study. Occasionally, however, studies require such enormous populations of organisms to demonstrate a background frequency or rate that the test data collected are routinely compared to a historical or cumulative data base. Certain types of animal studies, such as the heritable translocation assay (HTA) in mice or the SLRL assay in *Drosophila*, can be evaluated in this manner.

In the case of the mouse heritable translocation assay, a historical control is usually involved in the data evaluation, since the likelihood of finding a spontaneous translocation in the typical concurrent control populations is extremely low.[19] This is one of the few instances where historical control data is more important than concurrent control data. Some laboratories use historical values for bone marrow cytogenetic analysis and conduct only a small concurrent solvent control to check for unusual variation due to disease or other modifying parameters.

Similarly, in *Drosophila* testing for recessive lethal, a cumulative laboratory control value is often used in evaluating data, and only relatively small populations are used in concurrent controls. In this case, the laboratory cumulative control consists of many thousands of individuals and can be critical in data interpretation. Statistical tables giving the requirements for test sample size have been published for *Drosophila* studies using the average historical control frequency as a selection parameter.[44] The frequency of *Drosophila* SLRL in a given laboratory might be 0.20% based on accumulated control data from 50,000 flies. A single experiment is conducted with a small control consisting of 1000 flies and several concentrations of a test substance. Assume that the concurrent control frequency based on 1000 flies is 0.12% and that the recessive lethal frequency at the highest concentration of the test substance is 0.30%. While this frequency is almost threefold higher than the concurrent control, the cumula-

TABLE 4.8

Common Types of Control studies employed in Toxicology

Untreated negative control	Solvent vehicle control	Historical negative control	Positive control
Consists of the assay system without test agent or solvent/vehicle added. Defines the inherent background level of effects in the test. Gives a baseline from which test data and solvent/vehicle data can be interpreted. Is not essential in a minimal test, since the solvent control provides a baseline for interpretation. In tests that use no solvent/vehicle, however, this is an essential component.	A test of the solvent or vehicle used in the preparation and administration of the test substance. If a solution is obtained, the control is designated solvent control. If the test material is not solubilized, the control is referred to as a vehicle control. This control provides the baseline to which the test data are compared for significant effects. Any effects of the solvent/vehicle are assumed to be corrected for in this control.	A composite of previous control data collected from similar type studies. The historical control is a valuable aid in putting data into perspective, especially when the sample sizes for the concurrent control are relatively small. This control also permits variations in background levels due to environmental or temporal cyclic shifts to be compensated for in data analysis. Is not essential in a minimal test but may have critical value in certain types of biological assays.	Consists of conducting the assay with a substance known by previous experience to produce an effect in the test. Is not generally used as a baseline for interpretation of test substance data. Is not generally considered essential in a minimal test in which an adequate solvent vehicle control has been included. May be used to assess the ability of the investigator to produce and interpret effects in the test system. Control for *in vitro* activation activity.

tive control results might be viewed as eliminating the 0.30% value as a positive effect. Without the benefit of the historical data and the experience of what types of variation the controls show with small populations, data may occasionally be incorrectly evaluated.

Although not always conducted, positive control tests perform several functions in toxicology testing. The following points outline the type of information obtainable for positive controls:

1. They demonstrate that the test system responds to an agent previously shown to produce an effect and is functioning as expected.
2. They demonstrate the capacity of the investigator to identify and score the test system end point.
3. They can be used to demonstrate the reproducibility of the test system over an extended period. Significant temporal variations in response to a fixed concentration of a known control agent may signify problems with the test system or cyclical responses.

Valid tests can be conducted without the use of a concurrent positive control test for each test run. In fact, most animal-based toxicologic testing is performed without such controls. All of the roles defined above for positive controls can be evaluated by careful analysis of negative or solvent/vehicle control data. The proper functioning of a test system can be evaluated by examination of the viability, growth rate, and physiologic characteristics of the test organism coupled with a measurement of the frequency with which the end point occurs spontaneously. Likewise, an investigator's expertise can be determined through evaluation of scored spontaneous events.

In genetic toxicology tests involving bacteria, the inclusion of positive control values takes on more significance because it is not practical to examine the test organisms directly, and their capacity to respond to known genotoxic agents is the most reliable method for quality control. Frequently, in microbial or mammalian *in vitro* cell culture testing, an activation (S9 mix) system is employed for the biotransformation of nonactive molecules to their biologically active intermediate. The proper functioning of this activation system can be measured by the use of chemicals which produce positive responses following enzymatic biotransformation. Thus, positive controls for activation studies serve an extra function, that of quality-control determination of the S9 activation system. In fact, when substances are evaluated in a particular test both with and without an S9 activation system, it is necessary to include only the activation positive control to satisfy all of the functions of positive controls. Information from the nonactivation positive control is essentially redundant.

Concentration selection for positive control compounds is important. Ideally, the best concentration is one sufficiently high to show a consistently significant response, but low enough so that suboptimal test conditions will be detected by a loss or reduction of activity.

Compound selection for positive controls is also important. Similarity in structure to the test substance is advantageous, although often not possible, especially in large-scale chemical screening. Demonstrated activity with a single positive control of one chemical class cannot be assumed to validate the test system for chemicals of all classes.

Approaches to Data Analysis and Interpretation

Primary Evaluation

Each type of test system is amenable to some type of data analysis. Most are evaluated mathematically to determine the statistical significance of the results. A review of statistical analyses for short-term tests is given by Ehrenberg in the *Handbook of Mutagenicity Test Procedures* (see Appendix C). Those tests not subject to critical statistical analysis are generally evaluated by preestablished operational criteria. This usually involves setting a level below which the results are considered negative and above which they are called positive.

At the present time, very few universally accepted cutoff levels have been established. Most are based on laboratory experience or personal communication between investigators.[15] It is unlikely that such levels will be set in the near future. A careful review of published data and protocols such as those derived from the Gene–Tox program, ICPEMC, and the collaborative studies at the National Cancer Institute will first be necessary.

Interpretive Evaluation

The second level of data evaluation requires that the results of a test be used to describe the biological accuracy of a test substance. Is the compound a mutagen or a clastogen? Is there likely to be a health hazard following exposure to the material? This level of data analysis is even more elusive than the primary evaluation, mainly because there is no experience base to use as a reference. Therefore, most of the approaches related to interpretive evaluation are quantitative in nature and use a wide variety of modifiers such as potential, or suspect.

The three most common methods for evaluating results from screening batteries are:

1. The Decision Tree method in which a positive or negative result from a test leads eventually to a conclusion or categorization of a chemical. An example of the decision tree method for Ames test data is shown in Figure 4.6.

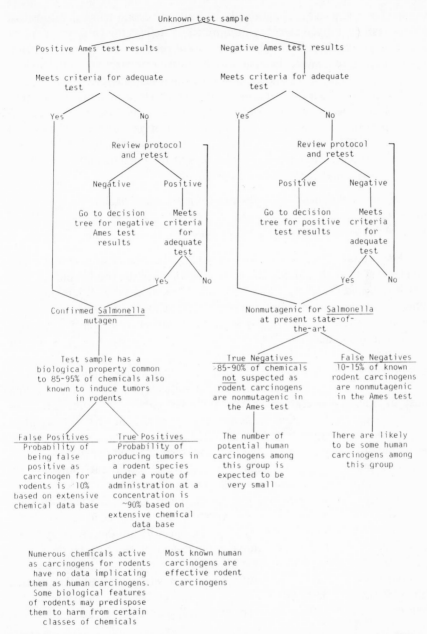

FIGURE 4.6. Ames test interpretation. This Decision Tree represents one manner in which test data may be interpreted. This chart attempts to define the probable consequences of each response. From Brusick, D. J.: In *Carcinogens: Identification and Mechanisms of Action* (A. C. Griffin and C. R. Shaw, eds.), Raven Press, New York, 1979.

2. The Critical Mass approach, in which a certain number of positive tests in a given category automatically defines the category of a test material regardless of the spectrum of response or the concentration needed to achieve the response. Negative responses are not contributory unless the results are negative in all tests conducted.
3. The Weighted Contribution approach, in which each test of a battery has a specific weight. The weight is proportional to the contribution that the test has in defining a positive or negative effect. The overall evaluation produces a summary test score which defines a test substance. This method is very useful for mixed results, since both positive and negative data can be handled.

This last method appears to be the most practical approach.

A Model System for the Weighted Contribution Approach to Data Evaluation

The primary questions in the practical application of genetic toxicology are (1) what tests should be performed; (2) what are the results and what do they mean; and (3) what, if anything, should be done next? While no single genetic tests can evaluate all types of possible genotoxicity, a variety of tests are available that, when employed in combination, offer the potential of decision making based on sound scientific grounds. Because it is almost certain that the test battery approach to genetic screening is going to be recommended for most situations, there is a need to coordinate the results of these tests into a unified evaluation classification scheme. This scheme must take into consideration all results of a test battery program, both positive and negative. Superficially, combining the results of several types of *in vitro* or submammalian tests into a unified evaluation classification scheme seems illogical, since many of the tests measure different genetic end points in phylogenetically different organisms. Conversely, without some method to coordinate and synthesize mixed results from a test battery, useful interpretation of the responses cannot be made. Often, the responses are synthesized on an intuitive basis by experts in the field of genetic toxicology; but without considerable expertise in the specifics of the assay systems, most individuals could not develop meaningful conclusions unless the results within the test battery were unanimous in either the positive or negative direction. Since unanimous results are not always obtained, the relative values of positive or negative responses must be weighed as a function of their contribution to the assessment of overall genotoxic potential.

Consistency of Test Battery Responses

Most practicing genetic toxicologists are aware that when a battery is constructed and the tests performed it is possible that the responses will be mixed. While disconcerting to the investigator, these responses can be predicted as shown in Figure 4.2. The rationale of the battery approach to testing is that the series of tests chosen will collectively identify a high proportion of true genotoxic agents without a high error rate. As illustrated in Figure 4.2, a few genotoxic agents are likely to be missed and a few nonmutagens will be incorrectly identified by an assay. The model predicts a significant amount of redundancy in the responses, but accounts for the occasional genotoxic compound identified by only a single test system. The problem with mixed results is how to sift through the responses and pull together a comprehensive evaluation summary for the test material. One approach is to assign to each assay a value based on its contribution, either positively or negatively, to an assessment of genotoxicity. The assignment of a value must take into account the following:

1. The type of end point measured by the test and its presumed relationship to the development of chronic toxicity *in vivo* (including mutagenesis and carcinogenesis).
2. The phylogenetic relationship of the test organism to mammalian species.
3. The reported reproducibility of the test system within and between laboratories.
4. The published data base supporting the utility of the test system to detect a broad range of chemical classes.
5. The susceptibility of the test to incorrect designations for genotoxicity resulting from testing artifacts or anomalous nonspecific responses.
6. The qualitative similarity between the test system metabolic or microsome bioactivation system and the *in vivo* bioactivation mechanisms in mammals.
7. The resolving power of the test system, including the strength of the data analysis methods used with the assay.

From these criteria, a table of assay values has been developed (Table 4.9). The tests listed in the table are among those which have been proposed as screening methods for animal mutagens and carcinogens.[2,3,7,12,23,28,31,33,34,37,41,43] They measure a variety of end points in target organisms including bacteria, insects, and cultured mammalian cells. Several restrictions are identified in the footnotes that specify the general criteria that must be met to consider an assay reliable.

The values presented in Table 4.9 for positive and negative responses range from −5 to +10. Assignment of the specific values was arbitrary with +10 as a maximum. The greatest negative value represents the test and test conditions that provide the most powerful indication of a lack of genotoxicity. Values between these two extremes are weighted propor-

TABLE 4.9

Score Table for Short-Term Test Results Using the Weighted Contribution Method

Test procedure[a]	Positive response[b]		Negative response[b]	
	Without activation	With activation	Without activation	With activation
Salmonella, Ames method	+6	+5	−2	−4
Gene mutation, cultured mammalian cells	+8	+7	−4	−5
Morphologic transformation *in vitro*	+10	+9	−4	−5
In vitro SCE induction	+3	+2	−1	−3
In vitro aberration induction	+4	+3	−2	−3
Drosophila SLRL	NA	+6	NA	−4
UDS in cultured mammalian cells[c]	+3	+2	−1	−3

Criteria for Negative Test Data

1. The lack of a positive response must be reproducible in chronologically independent tests using the same procedure.
2. The test material must be examined under test conditions demonstrating some evidence of biological activity such as cytotoxicity.
3. If the compound is known to be metabolized *in vivo*, an *in vitro* activation system must be employed in the evaluation. Mutagenic evaluation of any known mammalian metabolites in conjunction with the *in vitro* metabolism is recommended.
4. The ccompound should not be closely related to chemicals known to produce false-negative responses.
5. The test data should not differ significantly from a well-defined spontaneous mutation range.

Criteria for Positive Test Data

1. A positive response must be directly attributable to a genetic lesion at the DNA level.
2. A positive response must be significantly different from a well-defined spontaneous background.
3. A positive response should be dose related and shown to be reproducible in chronologically independent tests using the same procedure.
4. A positive response must be established as compound induced and not an artifact of the test conditions.
5. The genetically active chemical must be similar in nature to the form of the chemical found *in vivo*.
6. The compound is not closely related to chemicals known to produce false positive responses.

[a] Study design must include dose selection criteria, suitable controls, and provisions for multiple doses.
[b] Based upon criteria and analysis consistent with those given below.
[c] Only tests employing autoradiographic methods should be evaluated.

tionally according to the criteria listed above. Positive responses are obviously given significantly greater weight than negative ones, since negative results could mean either a lack of potential or a lack of detectability by the assay.

The differential weighting of results with or without S9 mix is predicated on the assumption that if a substance is active without the necessity for microsomal enzyme activation, the resonse is likely to be species independent. An effect dependent upon S9 mix activation, on the other hand, may be species restricted and not amenable to generalized extrapolation. For negative responses, however, a greater negative value is assigned to tests employing an activation system. The presence of such a system suggests that neither the parental molecule nor the microsomally produced breakdown products have detectable activity.

Table 4.10 is an attempt to bring a consideration of potency into the scoring system. Because of the diversity in end points measured, spontaneous rates for the detected events, and methods of scoring responses in the tests, a comparison of the absolute values is not feasible. A reasonable method for incorporating potency appears to be measurement of the lowest test concentration producing a significant increase over the control. This value is designated as the lowest positive concentration reported (LPCR). Another value, the highest negative concentration reported (HNCR), defines the highest tested negative concentration. The HNCR may be limited by toxicity or a preset maximum applicable concentration. The values listed in Table 4.9 under positive and negative responses can then be factored for potency into the weighting system presented in Table 4.10.

The product of the test score and the concentration score gives a total score (TS) for each test of the battery. The TS for each test will be either positive or negative. The algebraic sum represents the activity score (AS)

TABLE 4.10

Test Concentration Factors for Short-Term Test Score Table Using the Weighted Contribution Method

LPCR[a] or HNCR[b] converted to μg/unit[d] concentration	Concentration score point factor[c]	
	Positive response	Negative response
≤ 1.0	10	1
$> 1.0 \to 5$	9	1
$> 6 \to 10$	8	1
$> 11 \to 25$	7	1
$> 26 \to 50$	6	1
$> 51 \to 100$	5	2
$> 101 \to 500$	4	2
$> 501 \to 1000$	3	2
$> 1001 \to 5000$	2	2
> 5000	1	2

[a] LPCR, lowest positive concentration reported.
[b] HNCR, highest negative concentration reported.
[c] This factor is multiplied by the individual test score results obtained from Table 4.9.
[d] μg/unit, concentration in micrograms per milliliter or per plate.

NAME OF TEST SAMPLE:

	Column A If Positive	Column B If Negative	Column C	Column D
Name of Test	TV^a $LPCR^b$ [ ]	TV^a $HNCR^b$ [ ]	$Total^c$ Score	MP^d
2				
3				
4				
5				
6				
7				
8				

$$AS^e _____ MP _____$$

$$\frac{AS}{MP} \times 100 = _____ = Effect^f\ Category$$

FIGURE 4.7. Score form. [   ], Concentration score point from Table 4.10. ^{a} Table 4.9 value. ^{b} LPCR and HNCR from Table 4.10. ^{c} Maintain plus or minus sign in calculations. ^{d} MP, Maximum positive score; calculated by converting each minus score to a plus score and using the nonactivation column of Table 4.9 at the same concentration level at which the negative results were found (or 1000 μg/unit, whichever is lower). Then calculate a product score using the concentration factors in Table 4.10 (plus column) and enter in column C. These values are all added. ^{e} AS, Activity score, calculated as the algebraic sum of all scores in column C. ^{f} The ratio of the sum of scores in column C divided by the sum of scores in column D × 100.

for the compound in the battery of tests to which it has been subjected (Figure 4.7).

The next step in the scoring approach is to assign an effect definition to the AS. The effect definitions are shown in Table 4.11 and are calculated in the following manner.

1. A maximum genotoxic effect is calculated by taking each assay employed to evaluate the test material and calculating the TS product assuming it was positive without activation at a potency level equivalent to the concentration tested (e.g., a concentration of 250 μg/ml would have a potency value of 2 as a negative response but 4 when changed to a positive response.

2. This value, maximum positive (MP) total, is divided into the AS for the test material to obtain an index (expressed as a percent) of what portion of the maximum genotoxic effect was obtained in the evaluation.

3. The percent value is then categorized using the effect definition table (Table 4.11).

4. Each category in Table 4.11 defines the presumed genotoxic potential of the test substance and leads to an action response (Table 4.12).

TABLE 4.11
Effect Table

Percent maximum positive	Classification	Category
	Potent genotoxic agent	1
60%		
	Genotoxic agent	2
30%		
	Suspect genotoxic agent	3
10%		
	Insufficient response to categorize the agent as genotoxic	4

This scoring scheme, then, takes the actual test results from a multitest battery through to a specific action recommendation based on an index of the maximum genotoxic effect for that group of assays. This approach, or modified version of it, should facilitate the interpretation and decision-making processes which follow the actual testing program. In other words, the purpose of this type of scoring system is the use of multitest data to define

TABLE 4.12

Action Table[a]

Effect table category	% MP	Recommended follow-up action[b]
1	≥61	The test material must be considered to be a probable human mutagen and/or carcinogen. The sample would probably produce tumors in the standard rodent carcinogenicity bioassay. Further testing in risk assessment assays is not advisable unless the potential market value of the material is exceptional or it would be considered highly beneficial to humans. Human exposure to the material should be restricted.
2	31–60	If this value is based on a minimum of four assays, the sample should be subjected to a series of *in vivo* assays which are amenable to risk assessment determinations. Results of *in vivo* tests will be important, since category 2 responses may be considered as indicating the agent is a suspect human mutagen and/or carcinogen.
3	11–30	Additional testing is necessary to reach a conclusion regarding the genotoxicity of this agent. Consideration must be given to conducting the complete set of tests listed in Table 1. If the effect table category of 3 is maintained following the complete battery, the sample should be subjected to risk assessment tests as defined under category 2. If human exposure is considered low and the effect definition score is close to 10%, category 4 actions may be considered sufficient.
4	≤10	There is sufficient data to stop testing if a miminum of five assays have been conducted. This is especially true if anticipated human exposure is low to moderate. If fewer than five assays have been conducted, serious consideration should be given to bringing the number to five before terminating further testing.

[a] Based on the effect definition from Table 4.11, these sets of action have been recommended.
[b] The recommendations must be considered in the context of the actual testing. The spectrum of tests employed, the type of substance tested, and its intended use must all be factored into an action. The actions above are intended as general guidelines and do not reflect any type of anticipated regulatory decision regarding use or control of the test material.

the genotoxic potential of an agent, and to recommend follow-up action. The actions are intended to be consistent with good toxicology, and not necessarily to coincide with regulatory decisions regarding control of marketed materials.

Two examples of how the weighted contribution method of data analysis works are shown in Table 4.13. Tris and Fyrol FR-2 are structurally very similar, yet their genotoxic activity is distinctly different as

TABLE 4.13

A Comparison of Tris and Fyrol FR-2 in the Proposed Weighted Contribution Scoring System[a]

Agent	Name of test	Positive		Negative		Total	MP	Comment
		TV	LPCR (µg)	TV	HNCR (µg)			
Tris[b]	Ames Salmonella	+5	1 (10)			+50	+60	$\dfrac{AS}{MP} \times 100\% = 88\%$
	Mouse lymphoma	+7	50 (6)			+42	+48	
	SCE	+2	10 (8)			+16	+24	
	In vitro aberrations	+3	20 (7)			+21	+28	Category 1
	In vitro transformation	+10	5 (9)			+90	+90	
	Drosophila SLRL	+6	2000 (2)			+12	+12	
						231	262	
Fyrol FR-2[c]	Ames Salmonella	+5	200 (4)			+20	+24	$\dfrac{AS}{MP} \times 100\% = 12\%$
	Mouse lymphoma			−5	140 (2)	−10	+32	
	SCE	+2	130 (4)			+8	+12	
	In vitro aberrations	+3	130 (4)			+12	+16	Category 3
	In vitro transformation			−4	600 (2)	−8	+30	
	Drosophila SLRL			−4	>500 (2)	−8	+6	
						14	120	

[a] Data obtained from reference 8.
[b] Tris(2,3-dibromopropyl) phosphate.
[c] Tris(1,3-dichloro-2-propyl) phosphate.

TABLE 4.14

Evaluation of Selected Chemicals (of Commercial Interest) in the Weighted Contribution Genetic Scoring System

Compound	Tests conducted[b]	Effect category (MP)[c]	Reported animal	
			Mutagen	Carcinogen
Saccharin	1–7	4 (<10)	No	Yes (?)
Tris	1–7	1 (88)	No	Yes
Fyrol FR-2	1–7	3 (12)	?[d]	Yes
Formaldehyde	1, 2, 4, 5, 6	1 (72)	?	?
Benzene	1, 2, 4, 6	4 (<10)	No	Yes (?)
Dimethylsulfoxide (DMSO)	1–7	4 (<10)	No	No
Natulan	1, 2, 3, 5, 6, 7	1 (66)	Yes	Yes
Nitrilotriacetic acid (NTA)[e]	1–7	4 (<10)	No	No (?)

[a] All data obtained from Litton Bionetics, Inc., with the exception of those for Natulan and the *Drosophila* results for formaldehyde and benzene.

[b] 1, Ames *Salmonella* assay (agar incorporation method); 2, gene mutation in cultured mammalian cells; 3, *in vitro* cell transformation; 4, *in vitro* chromosome aberration assay using cultured mammalian cells; 5, *Drosophila* (SLRL); 6, SCE analysis in cultured mammalian cells; 7, UDS in cultured mammalian cells.

[c] From Table 4.11.

[d] Results unknown or (?) questionable.

[e] Results supplied courtesy Proctor & Gamble Co. Technical Center.

is the resultant recommendation from Table 4.12. Fyrol FR-2 is a good example of mixed responses from a test battery. Tris has already been defined as an animal carcinogen in several rodent bioassays. The results for Fyrol FR-2 are not presently available; however, based on this scoring system the carcinogenic properties of Fyrol FR-2 would be expected to be minimal.

Several other chemicals were put through the proposed scoring model and are listed in Table 4.14 with their respective category. Most of the data for these comparisons were taken from Litton Bionetics, Inc., or the published literature. It is preferable to develop original data for comparisons rather than selecting values from published reports.

It is immediately obvious from this small sample that benzene, which is a suspected human carcinogen, would not be detected as a genotoxic agent in the test battery employed. It is also interesting that benzene is also not detected in the rodent chronic bioassay methods. Saccharin is not detected as a genotoxic agent by this system primarily because the reported positive effects occurred only at very high concentrations. Formaldehyde is predicted to be a mutagen/carcinogen using this approach. Data from *in vivo* studies should be available soon to compare with this assessment.

It is anticipated that other approaches have been or will be devised to deal with screening test data. More work in this area is needed before genetic toxicology can be applied routinely to the regulatory decision-making process.

REFERENCES

1. Ames, B. N., Durston, W. E., Yamasaki, E., and Lee, F. D.: Carcinogens are mutagens. *Proc. Natl. Acad. Sci. U.S.A.* **72**:979, 1973.
2. Ames, B. N., Durston, W. E., Yamasaki, E., and Lee, F. D.: Carcinogens are mutagens: A simple test system combining liver homogenates for activation and bacteria for detection. *Proc. Natl. Acad. Sci. U.S.A.* **70**:2281–2285, 1973.
3. Ames, B. N., McCann, J., and Yamasaki, E.: Methods for detecting carcinogens and mutagens with the *Salmonella*/microsome mutagenicity test. *Mutat. Res.* **31**:347–364, 1975.
4. Arni, P., Mantel, T., Deparade, E., and Muller, D.: Intrasanguine host-mediated assay with *Salmonella typhimurium*. *Mutat. Res.* **43**(3):291–307, 1977.
5. Berwald, Y., and Sachs, L.: *In vitro* transformation of normal cells to tumor cells by carcinogenic hydrocarbons. *J. Natl. Cancer Inst.* **35**:641, 1965.
6. Bridges, B. A.: Some general principles of mutagenicity screening and a possible framework for testing procedures. *Environ. Health Perspect.* **6**:221, 1973.
7. Bridges, B. A.: Short-term screening tests for carcinogens. A review. *Nature (London)* **261**:195–200, 1976.
8. Brusick, D., Matheson, D., Jagannath, D. R., Goode, S., Lebowitz, H., Reed, M., Roy, G., and Benson, S.: A comparison of the genotoxic properties of tris(2,3-dibromo-

propyl)phosphate and tris(1,3-dichloro-2-propyl)phosphate in a battery of short-term bioassays. *J. Env. Pathol. Toxicol.* **3**:207–226, 1980.

9. Butterworth, B. E., and Golberg, l., eds.: *Strategies for Short-Term Testing for Mutagens/Carcinogens*, CRC Press, West Palm Beach, Fla., 1979.

10. Callen, D. F., and Philpot, R. M.: Cytochrome P-450 and the activation of promutagens in *Saccharomyces cerevisiae*. *Mutat. Res.* **45**:309–324, 1977.

11. Carrano, A. V., Thompson, L. H., Lindl, P. A., and Minkler, J. L.: Sister chromatid exchange as an indicator of mutagenesis. *Nature (London)* **271**:551–553, 1978.

12. Craig-Holmes, A. P., and Shaw, M. W.: Effects of six carcinogens on SCE frequency and cell kinetics in cultured human lymphocytes. *Mutat. Res.* **46**:375–384, 1977.

13. Deluca, J. G., Kaden, D. A., Krolewski, J., Skopek, T. R., and Thilly, W. G.: Comparative mutagenicity of ICR-191 to *S. typhimurium* and diploid human lymphoblasts. *Mutat. Res.* **46**:11–18, 1977.

14. de Serres, F., Fouts, J. R., Bend, J. R., and Philpot, eds.: *In Vitro Metabolic Activation in Mutagenesis Testing*, Elsevier/North-Holland, Amsterdam, 1976.

15. de Serres, F. J., and Shelby, M. D.: The *Salmonella* mutagenicity assay: Recommendations. *Science* **203**:563–565, 1979.

16. DiPaolo, J. A., Donovan, P. J., and Nelson, R. L.: *In vitro* transformation of hamster cells by polycyclic hydrocarbons: Factors influencing the number of cells transformed. *Nature (London) New Biol.* **230**:240, 1971.

17. Flamm, W. G.: A tier system approach to mutagen testing. *Mutat. Res.* **26**:329, 1974.

18. Gabridge, M. G., and Legator, M. S.: A host-mediated microbial assay for the detection of mutagenic compounds. *Proc. Soc. Exp. Biol. Med.* **130**:831–834, 1969.

19. Generoso, W. M., Cain, K. T., Huff, S. W., and Gosslee, D. G.: Heritable translocation test in mice. In *Chemical Mutagens: Principles and Methods for Their Detection*, Vol. 5 (A. Hollaender and F. J. de Serres, eds.), Plenum Press, New York, pp. 55–77, 1978.

20. Gletten, F., Weekes, U., and Brusick, D.: *In vitro* metabolic activation of chemical mutagens. I. Development of an *in vitro* mutagenicity assay using liver microsomal enzymes for the activation of dimethylnitrosamine to a mutagen. *Mutat. Res.* **28**:113–122, 1975.

21. Hollstein, M., McCann, J., Angelosanto, F., and Nichols, W.: Short-term tests for carcinogens and mutagens. *Mutat. Res.* **65**:133–226, 1979.

22. Huberman, E., and Sachs, L.: Cell-mediated mutagenesis with chemical carcinogens. *Int. J. Cancer* **13**:326, 1974.

23. Ishidate, M., Jr., and Odashima, S.: Chromosome tests with 134 compounds on Chinese hamster cells *in vitro*—A screening for chemical carcinogens. *Mutat. Res.* **48**:337–354, 1977.

24. Kakunaga, T.: The transformation of human diploid cells by chemical carcinogens. In *Origins of Human Cancer, Book C, Human Risk Assessment* (H. H. Hiatt, J. D. Watson, and J. A. Winstein, eds.), Cold Spring Harbor Laboratory, Cold Spring Harbor, N.Y., pp. 1537–1548, 1977.

25. Legator, M. S., Connor, T., and Stoeckel, M.: Detection of mutagenic activity of metronidazole and miridazole in body fluids of humans and mice. *Science* **188**:1118–1119, 1975.

26. Malling, H. V.: Dimethylnitrosamine formation of mutagenic compounds by interaction with mouse liver microsomes. *Mutat. Res.* **13**:425, 1971.

27. Malling, H. V., and Frantz, C. N.: *In vitro* versus *in vivo* metabolic activation of mutagens. *Environ. Health Perspect.* **6**:71–82, 1973.

28. McCann, J., Choi, E., Yamasaki, E., and Ames, B. N.: Detection of carcinogens as mutagens in the *Salmonella*/microsome test: Assay of 300 chemicals. *Proc. Natl. Acad. Sci. U.S.A.* **72**:5135–5139, 1975.

29. Milo, G., and DiPaolo, J.: Neoplastic transformation of human diploid cells *in vitro* after chemical carcinogen treatment. *Nature (London)* **275**:130–132, 1978.

30. Mishra, N. K., and DiMayorca, G.: *In vitro* malignant transformation of cells by chemical carcinogens. *Biochim. Biophys. Acta*, **355**:205–219, 1974.

31. Perry, P., and Evans, H. J.: Cytological detection of mutagen-carcinogen exposure by sister chromatid exchange. *Nature (London)* **258**:121–125, 1975.

32. Pienta, R. J.: A hamster embryo cell model system for identifying carcinogens. In *Carcinogens: Identification and Mechanisms of Action* (A. C. Griffin and C. R. Shaw, eds.), Raven Press, New York, pp. 121–141, 1979.

33. Pienta, R. J., Poiley, J. R., and Lebherz, W. B., III; Morphological transformation of early passage golden Syrian hamster embryo cells derived from cryopreserved primary cultures as a reliable *in vitro* bioassay for identifying diverse carcinogens. *Int. J. Cancer* **19**:642–655, 1977.

34. Purchase, I. F. H., Longstaff, E., Ashby, J., Styles, J. A., Anderson, D., Lefevre, P. A., and Westwood, F. R.: Evaluation of six short term tests for detecting organic chemical carcinogens and recommendations for their use. *Nature (London)* **264**:624–627, 1976.

35. Soares, E. R.: Genetic aspects of short-term testing and an appraisal of some *in vivo* mammalian test systems. In *Strategies for Short-Term Testing for Mutagens/Carcinogens* (B. E. Butterworth and L. Golberg, eds.), CRC Press, West Palm Beach, Fla., pp. 67–76, 1979.

36. Speck, W. T., Stein, A. B., and Rosenkranz, H. S.: Mutagenicity of metronidazole: Presence of several active metabolites in human urine. *J. Natl. Cancer Inst.* **56**:283–284, 1976.

37. Stitch, H. F., San, R. H. C., and Kawazoe, Y.: DNA repair synthesis in mammalian cells exposed to a series of oncogenic derivatives of 4-nitroquinoline 1-oxide. *Nature (London)* **229**:416–419, 1971.

38. Stout, D. L., and Becker, F. F.: Metabolism of 2-aminofluorene and 2-acetylaminofluorene to mutagens by rat hepatocyte nuclei. *Cancer Res.* **39**(4):1168–1173, 1979.

39. Syuki, A., and Sasaki, M.: Chromosome observations and sister chromatid exchanges in Chinese hamster cells exposed to various chemicals. *J. Natl. Cancer Inst.* **58**:1635–1640, 1977.

40. Tang, T., and Friedman, M. A.: Carcinogen activation by human liver enzymes in the Ames mutagenicity test. *Mutat. Res.* **46**:387–394, 1977.

41. Vogel, E.: The relation between mutation pattern and concentration by chemical mutagens in *Drosophila*. In *Screening Tests in Chemical Carcinogenesis* (R. Montesano, H. Bartsch, and L. Tomatis, eds.), IARC Scientific Publications No. 12, Lyons, 1976.

42. Vogel, E., and Sobels, F. H.: The function of *Drosophila* in genetic toxicology testing. In *Chemical Mutagens: Principles and Methods for Their Detection*, Vol. 4 (A. Hollaender, ed.), Plenum Press, New York, pp. 93–142, 1976.

43. Williams, G. M.: Detection of chemical carcinogens by unscheduled DNA synthesis in rat liver primary cell cultures. *Cancer Res.* **37**:1845–1851, 1977.

44. Würgler, F. E., Graf, U., and Berchtold, W.: Statistical problems connected with the sex-linked recessive lethal test in *Drosophila melanogaster*. I. The use of the Kastenbaum-Bowman test. *Arch. Genet.* **48**:158, 1975.

CHAPTER 5

Genetic Risk Estimation

INTRODUCTION

The state of the art in risk estimation for genotoxic effects in humans was probably best summarized by Newcombe:

> . . . little is known about the extent of the effect on health which a given increase in mutation rate would cause in man. This is true because of a lack of certainty about the amount of ill health that is maintained in the population by the pressure of recurrent natural mutations. Without such knowledge, the importance of exposures to known mutagens is difficult to assess quantitatively with a view to setting reasonable standards for the protection of the human gene pool.[17]

In other words, until the precise human genetic load baseline is determined, it will be difficult to place into perspective any predicted increase in mutations due to environmental mutagens. Many factors that would be critical to a risk assessment, such as the relative contribution of different types of genetic lesions to the risk estimate (e.g., dominant mutations, recessive mutations, chromosomal aberrations), are not well appreciated at the present time. It has been estimated that 1–2% of all liveborn humans are affected by genetic alterations resulting in hereditary disease.[4] The distribution of the estimated human genetic burden among the various classes of genetic lesions is shown in Table 5.1. Additional unknown contributions in the way of polygenic disorders also affect the human population. Distinguishing how many of these genetic lesions are the result of new spontaneous or environmentally induced mutations is extremely difficult. Because of the expense and time involved in monitoring all newborns for chromosomal and sentinel autosomal dominant mutations, the use of human epidemiology will not contribute significantly to risk evaluation procedures.

TABLE 5.1
Current Estimate of the Human Genetic Burden[a,b]

Genetic basis	Affected individuals per thousand
Chromosome abnormalities	6.86
Autosomal dominant gene mutations	1.85–2.64
Autosomal recessive gene mutations	2.23–2.54
Sex-linked gene mutations	0.78–1.99
TOTAL	11.72–14.03

[a] An additional 10–30% of congenital malformations and other complex disorders are believed to be the result of genetic alterations.
[b] From Committee on Mutagenicity of Chemicals in Food Consumer Products and the Environment. British Department of Health and Social Security Guidelines for Testing of Chemicals for Mutagenicity, 1979.

Although direct evidence for environmental induction of mutation in humans has not been demonstrated, the ability of specific chemicals to produce mutations in rodent species has been clearly demonstrated.[19] As in other areas of toxicology, rodent species are considered to be adequate models for humans in genetic evaluation. Genetic toxicology also relies extensively on several associated disciplines of toxicology such as pharmacology, physiology, biochemistry, and biometry for the development of information needed for risk estimates. This will be evident in the subsequent discussion of approaches to risk assessment.

DEFINITION OF RISK

Genetic risk can be viewed as the product of two factors: the inherent or intrinsic mutagenic potency of the compound being evaluated, and the ability of the compound to express any inherent activity in the DNA of target cells under normal exposure conditions (Figure 5.1). If either factor is zero, then regardless of the value of the other, the risk product will be zero. While it is not often that either factor is zero, some chemicals have values so close to zero that available test systems fail to detect effects. The units or methods of risk expression are variable.[22] Some common approaches to expression of risk are (1) a multiple of the dose required to double the normal background rate for the genetic end point being measured (the doubling dose method); (2) the number of new mutations or affected individuals resulting from a given dose. This method is also identified as the direct method and requires linear extrapolation from high test doses to low environmental exposures; (3) a function of or comparison to an equivalent dose of ionizing radiation such as defined by the REC (roentgen-equivalent-chemical) concept; or (4) a combination of the direct method and *in vitro* dosimetry.

An estimate of risk will probably require knowledge of the type of genotoxic alteration induced. Risk estimates based on data from tests which measure only a single end point, such as chromosomal damage or UDS, will not give adequate information relative to the risk from all classes of induced events. Species differences in pharmacodynamics, metabolism, and DNA repair capabilities must also have an impact on the risk estimate.

The risk from ionizing radiation, which by comparison with chemical risk is simple, is still subject to new interpretation and much scientific debate. It appears that earlier estimates of safe exposure levels for humans may have been too high. Accurate quantitative measurements of genetic risk from chemicals may not be possible with current rodent testing methods because of the severe limitations associated with definitively detecting point mutations. Risk estimates made with current methods must rely upon assumptions where data does not exist or cannot be obtained. Using these assumptions, it may be possible to distinguish between relative degrees of risk by maximizing the measured effect and comparing the maximized case to an established effect from a similar mutagen.[21] By using the present test methods and information derived from biochemical and metabolic studies, risks defined as low, moderate, or high can probably be estimated and safely extrapolated to humans. Such general categories are of potentially great value in decision making where several alternatives are available.

Several approaches to a unified concept of genetic risk have been proposed. The most notable and controversial was that of Abrahamson *et al.*[1] The hypothesis, designated ABCW, held that mutation rates per cell were a function of DNA content per haploid genome; and when plotted, the

$$N_1 \qquad \times \qquad N_2 \qquad = R \text{ (Risk)}$$

(Inherent genetic potential of chemical)   (Opportunity for hits on target molecule at normal exposure levels)

- N_1 can be detected using various assay systems on a qualitative basis; however, quantitation of genetic potential is difficult to measure except on a relative scale.

- N_2 cannot usually be quantitatively measured at the low environmental levels normally encountered. Even qualitative evidence for target molecule hit is difficult at low levels because of the magnitude of the populations that would be required.

- R can be roughly estimated at high dose levels using some type of N and N values. The meaning of R at high N_1 and N_2 levels is difficult to interpret and may not be directly extrapolatable to low levels.

FIGURE 5.1. Considerations involved in absolute quantitative risk assessment models.

TABLE 5.2

Comparison of Germ Cell and Somatic Cell Effects for Chemical Mutagens in Mammals Using Commonly Used *In Vivo* Assays

Compound detected as a germ cell mutagen	Assay[a,b]	Compound also detected in a somatic cell assay[b,c]
Ethylmethan sulfonate (EMS)	G1, G2, G4	S1, S2, S3
Methylmethane sulfonate	G1, G2, G4	S3
Triethylene melamine (TEM)	G1, G2, G3, G4	S1, S2, S3
Cyclophosphamide	G1, G2, G4	S1, S2, S3
Thio-tepa	G1, G3, G4	S1
Mitomycin C	G1, G3, G4	S1, S3
Myleran	G1, G4	S1
Colchicine	G1	S2
Trenimon	G1,	S1, S2

[a] *Germ cell assays:* G1, dominant lethal assay; G2, HTA; G3, specific locus assay; G4, spermhead abnormalities.

[b] Designations do not mean that unliested assays were negative but only that data were not available at this time.

[c] *Somatic cell assays:* S1, bone marrow cytogenetics; S2, micronucleus assay; S3, recessive mutation spot assay in mice.

rates showed a linear relationship across several different phylogenetic levels. The implications of this theory are significant, since it presumes a relatively simple extrapolation from microbes to man. This approach has been criticized because it does not take into consideration such parameters as the high levels of repetitive DNA normally found in mammalian cells, differential repair between species, and variable germ cell stage sensitivity.[23] The controversy has not been resolved, and it appears that this concept probably cannot be applied to all chemicals.

In the final analysis, risk also must be evaluated against societal or practical considerations such as risk/benefit, risk/risk, and economic impact. There will be unavoidable risks from naturally occurring mutagens or from materials impractical to remove from the environment. In these cases, it behooves society to know the relative, if not precise, impact of the risk. For example, it has been estimated that each human consumes about 10 tons of food (dry weight) by the age of 50, and some of this food contains naturally occurring or processing-derived mutagens.[10] It is not known what the impact, if any, of these agents is on our current genetic burden.[27].

Traditionally, estimates of genetic risk have been derived from studies of mutations in male and female gametes, since these cells provide the substrate for future generations. More recently, however, concern for somatic cell risk has increased. Somatic mutations, although not transmitted, may contribute significantly to the disease load, specifically as causative factors for malignancy, terata, and heart disease (Chapter 3). It may also be proposed that *in vivo* somatic cell mutations act as prescreens for germ cell risk since the estimation of risk would most likely be conservative. That is, a lack of mutation induction in somatic cells *in vivo* should indicate a corresponding lack of gamete risk. The argument made is that somatic cell evaluation would be a conservative surrogate for germ cell effects (Table 5.2).

Risk assessment in somatic and germ cells presents two different types of evaluations, each with unique problems (Table 5.3). The measurement of somatic cell risk appears to be somewhat easier than that of germ cells, at least at the current state of the art. The comparison outlined in Table 5.3 might also be interpreted as somatic cells being more reliable indicators of risk because a higher level of test resolution can be achieved. If this is true, then measurement of somatic cell risk may prove to be sufficient in making conservative risk/benefit decisions for chemicals. At the present time the

TABLE 5.3

A Comparison of Risk Assessment for Somatic and Germ Cells *in Vitro*

Comparison parameter	Somatic cells	Germ cells
Selective pressure on new mutants	Slight, since cells encounter only mitotic division	High, since meiotic cell division acts to screen out specific types of mutations such as chromosome effects
Barriers to mutagenic agents	Few barriers if agent is distributed systematically	Barriers similar to those for somatic cells plus blood/gonadal barriers (p. 121)
Sex differences	Minimal; some influence of hormones on metabolism in specific organs	Significant because of the different processes involved in gametogenesis between male and female mammals
Ease of detection of induced effects	Each dosed animal represents a population of exposed cells	Analysis generally requires large animal populations since each offspring represents a single exposed gamete analyzed
Size of exposed sample that can be scored	Extremely large population, generally 10^5 to 10^7 surviving cells scored	Generally small sample sizes; fewer than 10^5 animals in all cases
Type of genetic lesion detected	Chromosome aberrations and gene mutations both easily detected	Chromosome aberrations typically scored. Point mutations require extensive studies to detect with reliability

TABLE 5.4

Compounds of Suspected Human Genetic Risk Not
Detected in Standard Germ Cell Assays in Mammals

Compound	Detected as genetically active in	
	Germ cell assays	Somatic cell assays
Epichlorohydrin	No	S1
Benzo[a]pyrene	No	S1, S3
Benzene	No	S1
Urethane	No	S1
Hycanthone	No	S1, S3
Diethylnitrosamine	No	S3
Vinblastine sulfate	No	S1
Vinyl chloride	No	S1

[a] *Somatic cell assays:* S1, bone marrow cytogenetics; S2, micronucleus;
S3, somatic mutation.
[b] Designations do not mean that unlisted assays were negative but only
that data were not available at this time.

data base is insufficient to address this point in any great detail, and
attempts must be made to look at both somatic and germ cell effects.
However, it appears that few, if any, animal mutagens would be missed by
looking only at somatic cells, while some candidate mutagens would be
missed if only germ cell assays were employed (Table 5.4).

RISK ESTIMATION IN SOMATIC AND GERM CELLS

Somatic Cell Risk

Four methods have been described for somatic cell risk estimation;
three involve the use of animal models for direct measurement, and the
fourth is an indirect approach combining *in vitro* cell culture with *in vivo*
metabolism and chemical disposition. The first direct approach consists of
the measurement of chromosome aberrations in bone marrow or other
somatic cells of the exposed organism. These studies can be performed in
animal model experiments or by direct evaluation of human populations
through peripheral blood lymphocyte analysis. Problems abound, however,
when an attempt is made to interpret risk on the basis of chromosome
breaks, fragments, and other aberrations that are usually lethal to the
affected cell. Risk estimation must be based on the level of stable, heritable
aberrations observed. Therefore, analysis for sentinel aberrations known to

be essential in the formation of stable aberrations may be of some value. One such approach is to measure the frequency of dicentric (double centromere) chromosome aberrations. Dicentrics occur infrequently in mammals ($\sim\frac{1}{2000}$ metaphases) and are related to the formation of heritable translocation aberrations. This low spontaneous rate necessitates rather large experiments to detect chemically induced aberrations, but it is a feasible approach since slight changes can be measured. Measuring chromosome aberrations in bone marrow or other somatic cells is not a measure of gene mutations and cannot be considered an adequate method for risk evaluation if used in isolation because the relationship between concentration and type of mutation induced favors gene mutations at low doses.[29] The argument in favor of the cytogenetic analysis for risk estimation is that over half of the human genetic disease burden (number of affected newborns) is due to chromosome abnormalities (Table 5.1).

The second direct approach was developed by Dean and Senner in 1977 and is a method of mutation detection in Chinese hamsters. This method consists of exposing Chinese hamsters to mutagens, recovering tissues from the treated animals, preparing primary cell suspensions, and measuring the primary cells for forward mutation to 8-azaguanine or ouabain resistance induced in the DNA while cells resided in the animals. This method measures gene mutation induction and appears more sensitive than cytogenetic analysis, although the experience with this technique is limited to a single laboratory. Presumably, this method can be extended to an analysis for target tissue susceptibility; Dean and Senner have detected mutation induction in cells derived from the lung, liver, bladder, kidney, and stomach of Chinese hamsters.[8,9] This technique is exciting, for it is direct and requires only a small number of animals. Some technical difficulties involved with this technique remain to be solved; probably the most critical is developing the ability to increase viability and clonability of trypsinized primary cells. Application of this technique to human peripheral lymphocytes has been reported by Strauss and Albertini.[26] These investigators demonstrated increased mutation levels in humans exposed to chemotherapeutic agents.

The third approach to direct somatic cell analysis involves the measurement of coat color spots in F_1 mouse pups. The basic method (Figure 5.2), originally described by Russell and Major in 1957, consists of exposing pregnant female C57BL mice (mated to T males) near day $10\frac{1}{2}$ postconception.[19] The number of target cells (melanocyte precursor cells) per pup at this point is between 150 and 200. If it is assumed that the minimum number of scorable loci affecting coat color is four, then the mean number of cells at risk per litter of five pups is approximately 3500. If 50 pregnant females per dose group are used, the number of target cells at risk is 175,000. This represents a sizable number for data analysis, and a report by

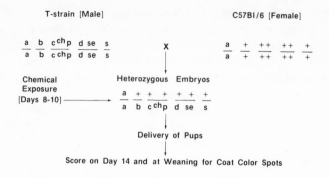

Description of Alleles

a   = agouti
b   = brown
c ch = chinchilla
p   = pink eyed dilution
d   = dilute
se = short ear
s   = spotted

FIGURE 5.2. Description of the basic protocol for measuring the induction of somatic mutations *in vivo*. Specific inbred strains are employed in this test. The target cells are the embryonic melanocytes carrying heterozygous alleles controlling coat color. The pups are exposed
in utero, and their hair is examined for spots (see Figure 8.4).

Russell indicates that the somatic test is more sensitive than the associated
specific locus germ cell test.[20] Comparisons with seven mutagens showed
that the dose required to demonstrate mutagenic activity was lower for
melanocyte precursors (somatic test) than for spermatogonia (germ cell
test); with three of the mutagens, hycanthone, benzo[a]pyrene, and DMN,
no mutants were detected in spermatogonial experiments (Table 5.5).
Russell has attributed the production of coat color spots to several types of
somatic lesions, including point mutation, small chromosomal deletions,
somatic recombination, and whole chromosome loss.[19] At the present time,
a small but growing data base has been accumulated for the somatic mutation coat color assay. The results are summarized in Table 5.6.

The target cell numbers associated with this assay are reasonably large,
and the test does not suffer from the technical limitations described for the
tissue culture analysis of Dean and Senner in Chinese hamsters. However,
differential placental transport of chemicals among species has been well
documented,[11] and uncertainty of adequate fetal exposure complicates
interpretation of negative data from the somatic spot test.

The fourth approach is an indirect method of somatic cell risk. This
integrated technique of somatic cell risk couples *in vitro* mutation induction
with DNA adduct formation. Alkylating agents as well as other reactive

molecules form covalent-bound DNA adducts at various sites which can be measured biochemically if the chemical is tagged with a radiolabel such as ^{14}C. Other techniques, using immunofluorescence detection systems, are very sensitive and can detect a single adduct per 10^6 nucleotides. Using these methods to measure the dose, the activity of an agent can then be defined as a ratio of

$$\frac{\text{Induced mutants}}{\text{Units of alkylated DNA bases}} = \text{Mutagenicity index (MI)}$$

Standard MI curves for specific reference chemicals of different adduct-forming classes can be developed for use in evaluating "unknown" samples. By dosing animals with the labeled "unknown" test substance, collecting somatic tissue, and calculating the number of DNA adducts produced per genome, the projected level of induced mutants per cell can then be calculated from a standard curve. This approach is a coordinated effort, and can be performed rapidly and with a high degree of sensitivity assuming that the test material is amenable to radio- or fluorescent-labeling and forms DNA adducts.

Cells from various suspect target tissues can be analyzed. One important aspect of DNA adduct formation identified by this method is that different alkylation or adduct sites are associated with different levels of

TABLE 5.5

A Comparison of the Sensitivity of Somatic and Germ Cells to Specific Locus Mutation Induction by Eight Agents[a]

| Chemical | Gene mutation response in | | Activity ratio[d] |
	Germ cells[b]	Somatic cells[c]	
EMS	+	+	NA[e]
Methylmethane sulfonate	+	−	NA
TEM	+	+	2.8
Mitomycin C	+	+	2.6
Benzo[a]pyrene	+?	+	5.0
Hycanthone	−	+	NA
Diethylnitrosamine	−	+	NA
Natulan	+	+	4.0

[a] Modified from Russell.[19]
[b] Effect based on data from spermatogonia or spermatozoa (whichever appeared more sensitive).
[c] Data from the mouse *in vivo* somatic coat color mutation assay.
[d] Effective germ cell dose/Effective somatic cell dose = activity ratio. Values >1 suggest greater sensitivity in the somatic cell test.
[e] NA, not applicable.

TABLE 5.6

Current Data Base of Mutagens Reported as Active in the *in Vitro* Mouse Recessive Spot Test[a]

| | Also active as a mammalian | |
Compound	Germ cell mutagen	Carcinogen
EMS	+	+
TEM	+	+
Mitomycin C	+	+
Benzo[a]pyrene	−	+
Hycanthone	−	+ (?)
Cyclophosphamide	+	+
Diethylnitrosamine	−	+
DMN	−	+
N-Nitroquinoline N-oxide	+	+
Acridine orange	−	−
X radiation	+	+
2-aminopurine	−	−
Procarbazine	+	+
Methylbenzimidazole-2-carbamate (MBC)	−	−
Trichloroethylene	−	+ (?)
N-Methy-N'-nitro-N-nitrosoquanidine	−	+

[a] Fahrig, R.: The mammalian spot test: a sensitive *in vivo* method for the detection of genetic alterations in somatic cells of mice. In *Chemical Mutagens*, Vol. 5 (A. Hollaender and F. J. de Serres, eds.), Plenum Press, New York, pp. 151–176, 1978.

induced mutations. For example, alkylation at the O-6 position of guanine is more likely to result in mutation than alkylation of the N-7 position.[14] This type of information is critical for quantitation of effects and must be employed in determining the potential genotoxic risk for a given agent.

A review of these and other methods for *in vivo* somatic and germ cell mutagenesis has been published by Malling and Valcovic.[15]

Germ Cell Risk

The primary goal of genetic toxicologic evaluation is to determine whether a chemical or mixture of chemicals might increase the level of genetic burden to future generations. Of concern are the female and male gametes and the stem cells which give rise to the gametes. The process of gametogenesis is different in male and female mammals (Figure 2.11), and as a result, the risk factors associated with germ cell exposure to chemical mutagens differ. Male gametes (sperm) are produced on a continuous basis over the reproductive lifespan of male mammals, while in the female, oogonia proceed through meiosis until the diplotene stage of the first meiotic prophase and remain arrested until ovulation. These fundamental differences in gamete formation profoundly influence the susceptibility of male and female genes to chemical mutagens.[3]

The accessibility of chemicals to maturing ova and sperm is of critical importance to the development of risk estimates, as is the clearance of exposed gametes. Male gametes turn over every 8–10 weeks, and mutant sperm are purged from the system, whereas the meiotically arrested female gametes remain *in situ* unless they are ovulated. This can be important when animals are exposed intermittently to mutagens. If maturing male gametes are exposed but not involved in fertilization, they will be cleared approximately every 8–10 weeks and new unexposed sperm will replace them. Oocytes, however, will not be lost until ovulation, and the majority remain *in situ* during numerous sequential exposures.

Biologically active molecules have the opportunity to form other interactions prior to reaching DNA in the sperm or ova, and it is the quantification of the effect of these factors which complicates the calculations that are part of risk assessment. The process of studying germ cell risk can be broken into five components:

1. Selection of the route of exposure and its impact on total body dose received.
2. The pharmacodynamics of the material and its metabolic fate in the test species.
3. Blood/gonad barriers and their effect on the target organ dose experienced.
4. Development of dosimetry data for the target molecule DNA.
5. Measurement of an actual genetic effect in the target cell of the intact organism.

Relationship of the Route of Exposure to Total Body Dose

Genetic toxicology studies have been conducted and reported using all routes of compound administration. Table 5.7 lists the most common methods of exposure and their limitations. The total dose received and its distribution varies considerably according to the route selected. For example, rats exposed to 500 ppm of a compound will be subjected to different total real doses according to the route of administration.

Using current toxicologic procedures, it is virtually impossible to precisely calculate total body dose for chemicals encountered under normal conditions. However, for the purpose of estimating risk, one can rank the relative risk associated with certain types of exposures. Table 5.8 shows the ranking based on the volume of a compound likely to be consumed and/or absorbed during a given period of time at a fixed concentration. While the relationships shown in the table were developed from rodent data, they apply to all mammalian species. This aspect of dosimetry is critical to risk methods not using a direct measurement of reaction products at the target molecule.

A similar level of uncertainty exists for target organ dose, since the route of administration affects the type of primery metabolism of the sample. The microsomal P-450 enzyme composition of lung tissue may differ from that of liver tissue and may assume the production of qualitatively or quantitatively different metabolites at the primary metabolic site. This difference could influence the target organ effect prior to liver activation.[7] Thus, route of exposure and portal of entry are important considerations in analyzing chemical fate *in vivo*.

Relationship of the Route of Exposure to Metabolism

The *in vivo* distribution and metabolic fate of a chemical is dependent to a certain extent on the route of exposure. For instance, oral dosing has the advantage of allowing for a dose of significant mass, but it can give misleading information if the test substance is acid labile, resistant to gut transport, susceptible to bacterial degradation, or becomes readily trapped by the excreta. As a result of oral administration, the test substance enters the hepatic portal system and is, of course, immediately subjected to the liver and its profound metabolic activity. Once leaving the hepatic circulation, the test substance is then exposed to the general circulation.

TABLE 5.7
Routes of Exposure and Their Limitations

Routes	Limitations

Oral
  Feed
  Drinking water
  Gastric intubation
	Must identify stability in feed or water as well as intake during exposure.
	Places stress on animals and gives agent by pulse rather than continuous dose.

Inhalation
  Respiratory
  Intratracheal insufflation
  Implantation
	Inhalation dose monitoring is technically difficult for certain substances, especially complex mixtures of many different substances. There can be a significant amount of ingestion of material carried in the hair during grooming after exposure.
	Particle size is important in defining the actual total body dose.

Dermal
  Skin painting
  Injection (subcutatneous)
  Intravenous
  Intraperitoneal
	Localized immunologic or irritation effects can cause problems; skin absorption must be identified to understand dose.

TABLE 5.8

The Relative Risk Resulting from Exposure to a Model
Compound[a]

Route of exposure[b]	Relative level of risk[c]
Inhalation (nonparticulate)	Highest
Diet (feed)	↓
Drinking water	
Dermal (skin contact)	Lowest

[a] The model compound produces optimal absorption under all conditions and is not metabolically altered.
[b] Concentrtion × time (e.g., 500 ppm for 24 hr).
[c] Based on a total body dose achieved following exposure to the specified concentration by various routes.

Intravenous dosing, on the other hand, has the advantage of direct introduction into the general circulation. A large volume of the test substance is usually not possible via intravenous injection because of its usual direct toxic effects on the experimental animal. Furthermore, in the case of lipophilic products, a suitable vehicle must be found, and this further complicates the dosing situation. Once these problems are overcome, however, the product is in the general circulation, and the most elucidative picture of specific organ metabolism and toxicity is usually revealed. Intraperitoneal injections are not that dissimilar to intravenous injection since the test substance quickly reaches the general circulation and gives the same pertinent information.

Dermal application or dosing has numerous problems that involve dermal trapping, evaporation, physical removal (e.g., licking or grooming), and poor absorption. There is known metabolic activity in the skin, and the pharmacological events may begin immediately upon application. It is probably fair to presume, however, that once transported into the general circulation the test substance proceeds as if given intravenously.

Inhalation dosing and the subsequent metabolic studies would probably have to be well controlled so that dermal exposure does not occur simultaneously. The lung is a proven metabolic organ, and this could provide initial alterations in the test substance. Here again, following transport across the lung and its initial metabolic impact, the test substance would be in the peripheral circulation for general metabolism.

Blood/Gonadal Barriers

Levels of chemicals in the bloodstream in the systemic circulating system can be used to approximate somatic cell exposure, but are not necessarily accurate for germ cell exposure since the gonads are somewhat protected from the general circulation by what are referred to as blood/gonadal barriers. The movement of nonelectrolytes from the body circulation to the testes, for example, is dependent on molecular size.[12] Molecules smaller than 3.6 Å can be transported readily, while larger molecules cannot. Lipid solubility is also important, and studies suggest this is a strong rate-limiting factor in penetration at physiologic pH.[12] There is minimal evidence in the testes indicating active transport, although certain antibiotics and metal ions appear to be concentrated in rete testis fluid to a greater extent than they are found free in plasma.

Penetration of chemicals into the ovary also has been studied.[11] The ovarian follicle is an avascular structure, and before a chemical can reach an oocyte located in a maturing follicle it must leave the vascular system, cross the thecal layer, and enter the follicular fluid.[11] Once in the follicular fluid, an agent moves passively into the oocyte. The same factors which affect transport across the blood/testis barrier (molecular size, lipid solubility, and ionization) also affect follicular transport in the ovary.

In general, then, gonadal structures restrict the permeability of many foreign compounds to the mature male and female gametes (spermatogonia, however, may lie outside the blood/testicular barrier). Little active or selective transport occurs. The major transport of chemicals depends on their lipid solubility and molecular size, and penetration appears to obey simple diffusion kinetics. As a result of this restriction of transport, the gonadal barriers probably reduce the risk of exposure of germ cells compared to somatic cells.

MOLECULAR DOSIMETRY OF CHEMICAL MUTAGENS

Dose is generally described as a function of concentration and time. However, this approach to dose calculation does not take into consideration several problems such as the nonlinearity of exposure to dose ratios, rapid loss or alteration of the molecule during exposure, and differences in metabolism as a function of concentration and length of exposure. The most accurate measure of dose is the quantification of reaction products as described in the section on somatic cell risk. Using alkylating agents as models, alkylated DNA bases can be defined as the unit of dose determination. This approach may not be useful for studies of certain mutagens, such as intercalating mutagens, which do not covalently bind to macromolecules. Lee reported the relationship between EMS dose in *Drosophila* and mice and ethylations per sperm cell to be linear.[13] Data from SLRL studies in *Drosophila* showed a linear relationship also existed between ethylations/nucleotide and percent mutation.[13] If mutation per locus per alkylation is roughly equivalent across species, then extrapolations across mammalian species should be possible. The ability to perform this type of dosimetry is almost unique to genetic toxicology, where the specific target molecule is known and can be examined with very sensitive methods. Figure 5.3 shows the ethylations/nucleotide relationship for recessive lethal mutation, dominant lethal mutations, and reciprocal translocations as derived by Lee.[13]

Two additional considerations are important in risk calculations. One is the effect of DNA repair processes on reaction products. It has been demonstrated that the efficiency of repair is dependent on the type of reaction product formed and the meiotic cell stage exposed. Many of the metabolic and repair capabilities are suppressed in mature gametes, making them somewhat variable in sensitivity depending on the nature of the chemical. Direct-acting alkylating agents, for example, are very effective mutagens for mature sperm but not so effective for premeiotic spermatogonia.[21,32] There are also quantitative differences in repair capacity between species which can alter the formation of reaction products (Figure 2.20). Another factor is target size. Eukaryotic organisms, mammals in particular, have a high content of repetitive DNA (nonfunctioning DNA) in their chromosomes.[15] Since the amount of DNA appears to influence the yield of mutation, dosimetry must be able to define an effective dose (i.e., number of reaction products per active gene) taking into account the amount of repetitive DNA sequences in a given species.

MEASURING THE GENOTOXIC EFFECT

Detection of an effect requires more effort than measuring reaction products since the reaction product : effect ratio is not 1 : 1. For example, a 1% SLRL frequency is generated by $\sim 1.2 \times 10^5$ nucleotide ethylations per sperm cell.[13] Most reaction products (DNA adducts) appear to be either repaired or located at noncritical sites in the genome. This necessitates that large-scale animal studies must be conducted to observe the rare biological expression of altered DNA as a mutation in an intact organism. As an example, the mouse specific locus assay can be conducted with a rather wide range of animals scored. Analysis of the results from the test is made according to the following calculations:

$$\text{Multiple of the spontaneous rate not excluded} = \frac{(\text{Upper } 95\% \text{ confidence limit observed/population scored}) - \text{spontaneous frequency}}{\text{spontaneous frequency}}$$

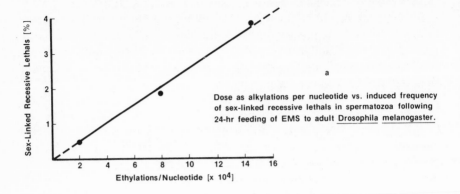

a

Dose as alkylations per nucleotide vs. induced frequency of sex-linked recessive lethals in spermatozoa following 24-hr feeding of EMS to adult Drosophila melanogaster.

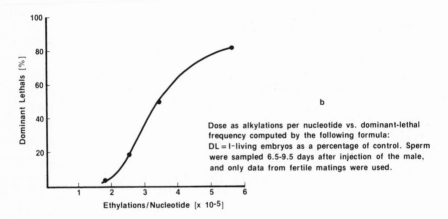

b

Dose as alkylations per nucleotide vs. dominant-lethal frequency computed by the following formula:
DL = I-living embryos as a percentage of control. Sperm were sampled 6.5-9.5 days after injection of the male, and only data from fertile matings were used.

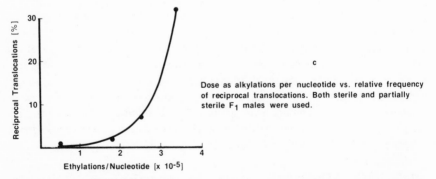

c

Dose as alkylations per nucleotide vs. relative frequency of reciprocal translocations. Both sterile and partially sterile F_1 males were used.

FIGURE 5.3. Relationship between the formation of reaction products and genetic effects in *Drosophila*. This type of data could be adapted to risk assessment if the alkylations/nucleotide versus mutation induction relationships hold across species lines. Adapted from Lee.[13]

If we assume that the results from a specific locus assay scoring 1000 F_1 progeny yielded no mutants, the experiment could exclude an effect which did not increase the spontaneous frequency by 67-fold:

$$\frac{(3.3/1000) - (28/531,000)}{28/531,000} = 67$$

However, about 12,000 F_1 progeny would have to be scored to exclude an effect not exceeding four times the spontaneous (assuming no mutants found). Similar discussions of numbers and data analysis are given for the mouse heritable translocation and *Drosophila* SLRL assays in Chapter 8. At the present time, the data base for the sex chromosome loss is insufficient to discuss in detail.

Genes play different roles in cell operations. Therefore, the specific site of a genetic lesion is critical, since the manner of expression of an altered gene depends on the role of the gene in the organism. Table 5.9 summarizes

TABLE 5.9
Location of Genetic Effects and Ability to Detect Effects in the Target Molecule[a]

Type of gene	Function	Ease of direct detection
Structural gene	Codes for the amino acid sequence in a protein (Figure 2.2)	Easily detected—forms the basis of most genetic assays used in routine screening
Operational genes Modifying gene	Modifies the protein structure after the primary polypeptide forms.	Can be measured by genetic tests such as those measuring temperature-sensitive mutations. However, not commonly detected in routine screening tests.
Architectural gene	Determines the organ in which enzyme expression will occur.	Extremely difficult to identify. Almost never detected in screening.
Temporal gene	Determines the time in development at which a structural gene is turned on or off.	Extremely difficult to detect. Almost never detected in screening.
Regulatory gene	Modifies the production of gene products in response to internal or external environments.	Can be detected in microbial systems, but is seldom used in screening.

[a] As modified from Malling and Valcovic.[15]

both the kinds of functions controlled by genes and the probability of directly detecting a point mutation induced in such a gene. Detection of altered structural genes is the usual method of mutation identification because the mutant phenotype can usually be readily identified (e.g., the requirement for nutrients in bacteria or hair color changes in mice). Detection of operational genes usually requires highly specialized research methods. However, mutations occurring in nonstructural genes could contribute substantially to the induced genetic burden.

It has been estimated that the haploid human genome contains about 6×10^9 nucleotides, or some 30,000–50,000 genes.[15] Among this number, the ratio of the various types of genes identified in Table 5.10 is not known, but a high proportion would be expected to be structural, again making these sites logical targets for routine screening. Because so much of the genome of a typical mammal is repetitive DNA, it has been estimated that the human chromosomal target appears to be 90% nontranscribed repetitive DNA and only 10% critical target DNA.[15]

The currently available animal tests for germ cell effect assessment are listed in Table 5.10. This is a relatively restrictive list which emphasizes the overall lack of good methods for direct risk estimation. *Drosophila* tests are included, but are considered by many to be inappropriate for chemical risk assessment because the test organism is an insect and may have sufficient differences in metabolism and gamete development to prevent reliable extrapolation. Other investigators, however, feel *Drosophila* tests are appropriate in this role.

TABLE 5.10

Available Germ Cell Risk Assessment Assays for Direct-Effect Measurement in the F_1 Generation

Test	Organism	End point measured	Reference
Specific locus assay	Mouse	Presumed gene mutations and small deletions	24
HTA[a]	Mouse, *Drosophila*	Balanced chromosome translocations	30
SLRL assay[a]	*Drosophila*	Presumed gene mutations, small deletions, balanced translocations, and X and Y chromosome loss	32
Sex chromosome loss assay	Mouse, *Drosophila*	Loss of X or Y chromosomes by nondisjunction	18

[a] Sample study designs given in Chapter 8.

TABLE 5.11

A Comparison of the Direct and Integrated Approaches to Risk Estimation Using Animal Models

Aspect	Direct	Integrated
Type of test	Specific locus assay in mice Heritable translocation in mice Sex chromosome loss in mice	Isolation of DNA from the target cell of an exposed animal and measurements made of DNA adducts
Advantages	Quantitation of genetic event per gene vs. dose Samples all stages of spermatogenesis Heritability unequivocally demonstrated	Precise dosimetry at target site is known, permitting quantitative comparisons between species Small numbers of animals required Can use male or female animals Approach is amenable to use in humans
Limitations	Lack of quantitative dosimetry at the target cell Sensitivity of the test is low unless very large populations are scored in the F_1 generation ($> 10^3$) Tests are almost exclusively limited to risk in male gamete Generally requires exposure to concentrations considerably higher than that encountered in the environment Minimal pharmacodynamic data collected It is not known if the loci are representative of all genes	Actual measurement of genetic event is not made Requires ability to uniquely identify that portion of the test molecule which reacts with the target site (e.g., alkylations per nucleotide) Mutagen index is based on model compounds such as EMS which may have a different type of covalent binding than the test substance

a From Valcovic.[28]

APPROACHES TO GERM CELL RISK USING ANIMAL MODELS

Applying all of the previously discussed parameters, germ cell risk estimation can be measured using two basic approaches: the direct method and an integrated one.[28] The two approaches are compared in Table 5.11; however, specific techniques and data analyses are likely to vary with the philosophy and background of the investigator. In the direct approach the dose, pharmacological barriers, metabolism, and effect are ignored as individual components and risk is assessed by the quantitative production of an effect (end point) in the intact organism as listed in Table 5.10. The integrated approach, on the other hand, attempts to define the actual dose received at the target site using biochemical or immunological methods and, by data extrapolation of the type developed by Lee, to estimate the level of effects that would be induced in the intact animal.[13] However, no actual effects are measured using this approach. The integrated approach relies heavily on exposure, pharmacodynamics, and metabolism. For instance, the target molecule is isolated and analyzed for quantification of alkylated sites. At the moment, neither risk assessment approach is validated to a point that one could be recommended over the other. Most current risk assessments have been conducted using the direct approach, the specific locus, or heritable translocation test systems. As more data of the type shown in Figure 5.3 is developed, it is likely that the indirect method will be used more often, as the indirect method appears to be the more sensitive and cost-effective approach to risk estimation but will require considerable development and validation efforts over the next few years.

EXTRAPOLATING DATA FROM SHORT-TERM TESTS TO HUMANS

Using test data from all available sources, the genetic toxicologist must try to extrapolate the results of tests to the human experience. The fact that the structure and function of the target molecule, DNA, is essentially the same in all organisms permits a reasonable expectation of success. The primary complicating factors are summarized below in an illustration with a carcinogen risk assessment.

1. Most tests, especially *in vivo* ones, are conducted with small test populations at high exposure levels. The proper extrapolation of results down to the typical environmental exposure levels for large populations is disturbingly uncertain (Figure 5.4), and the estimate of human risk depends on the statistical model employed. The uncertainty of risk estimation at low dose levels is clearly demonstrated in Table 5.12, where the estimated increase in cancer incidence from exposure to saccharin was from 0.0007 to 3640 cases per 50 million population per year, depending on the statistical model used.

2. There are likely to be substantial pharmacodynamic differences between species, and these can lead to errors when judging susceptibility. For example, thalidomide, the human teratogen, is considerably more effective in humans than in the typical rodent model test system.[31] Thus, preliminary risk estimations using rodent species would have erred in the wrong direction.

3. Basic differences in physiology and organ pharmacodynamics may also play an important role in extrapolation of risk. For example, the intrinsic capacity of DNA repair systems is different in rodent species from

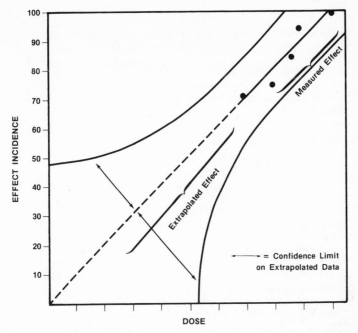

FIGURE 5.4. The assumptions and considerations involved in the linear one-hit Model for data extrapolation. This model assumes that the effects measured at high dose levels can be linearly extrapolated to zero dose. This may not be true for all genotoxic substances, especially those that act through repairable DNA lesions (see Figure 5.5).

that of humans (Figure 2.20); repair capacity can be extremely important when considering extrapolation of effects at usage levels where most lesions may be corrected. Pulmonary physiology, immunocompetence, placental permeability, renal function, and hormonal variability are also different, and as a result may directly influence the expression of an effect in various mammalian species.

4. The degree of potency must be considered in extrapolations of animal data. Potency is the intrinsic activity of a mutagen, and is a product of the chemical's DNA affinity and site specificity.

EXTRAPOLATION OF *IN VITRO* DATA DIRECTLY TO *IN VIVO* RESPONSES

As indicated in Chapter 2, numerous attempts have been made to establish a relationship between the mutagenic activity of carcinogens *in vitro* and their carcinogenic activity *in vivo* (typically in rodent bioassays). With the *Salmonella* system, Meselson and Russell[16] were able to

TABLE 5.12

Estimated Human Risks from Saccharin Ingestion at 0.12 g/day[a]

Method of high- to low-dose extrapolation	Lifetime cases/ million exposed	Cases per 50 million/year
Rat dose adjusted to human dose by surface area rule		
Single-hit model	1200	840
Multistage model (with quadratic term)	5	3.5
Multihit model	0.001	0.0007
Mantel–Bryan probit model	450	315
Rat dose adjusted to human dose by mg/kg/day equivalence		
Single-hit model	210	147
Multihit model	0.001	0.0007
Mantel–Bryan probit model	21	14.7
Rat dose adjusted to human dose by mg/kg lifetime equivalence		
Single-hit model	5200	3640
Multihit model	0.001	0.0007
Mantel–Bryan probit model	4200	2940

[a] From NRC/NAS Report on Saccharin.[6]

demonstrate a rough relationship between their *Salmonella* data and carcinogenicity. Similar results have been reported from Ames' laboratory, and a major effort to expand this data base is in progress. A second comparative report using the *in vitro* gene mutation system in L5178Y mouse lymphoma cells as the mutagenic parameter was conducted by Clive *et al.*[5] His data also demonstrated a potency correlation between *in vitro* and *in vivo* responses.

These studies have raised hopes that categories of carcinogenic potency could be determined by *in vitro* mutagenesis assays. Critics quickly pointed out several potential problems with this approach.

1. Selection of the specific target cell for mutagenic analysis can greatly influence the reported potency of the chemicals being compared. For example, *Salmonella* mutants TA-1538 and TA-98 are both derived from 3052, yet for a given chemical mutagen significantly quantitative differences between strains may be obtained. No rules have been formulated to decide which should be used in developing the correlation.

2. If a target organism is selected that gives apparent correlation under one set of test conditions, the correlation may be significantly altered by slight modifications in the study design. This suggests that a multitude of different study designs would have to be used to determine which gives the proper correlation, and adoption of uniform study designs probably would not yield the proper consistent relationships.

3. *In vitro* metabolic activation systems most often employ S9 homogenate preparations. These preparations are prepared from animals pretreated with inducing agents and do not accurately reflect normal *in vivo* metabolic processes. (See Appendix A for S9 homogenate preparation methods.) Choice of species, inducers, and levels of S9 employed in the *in vitro* assays can modify the degree of response *in vitro* and significantly modulate the correlation of a chemical within a given test system.[2] The use of intact hepatocytes, which probably give more accurate activation responses, will give yet a different set of correlations.

4. If data points from noncarcinogenic mutagens and mutagenic noncarcinogens are added to the comparisons shown in Figure 3.10, the picture becomes substantially more complex and less convincing.

Because of the ease with which *in vitro* potency can be modified via the methods described above, it is unlikely that an exact *in vivo* response can be predicted with accuracy. A general relationship for some types of chemicals (those which do not require activation) may be observed between *in vitro* and *in vivo* responses, but the only reliable methods presently available for assessing *in vivo* potency are with animal models.

Threshold

Discussions of risk estimation and extrapolation invariably lead to the subject of threshold levels for toxic effects. Certainly in the area of reversible toxicity, "no recognizable effect" levels can be established. The same cannot be confidently stated for irreversible effects such as mutations and oncogenic transformation. The types of animal bioassays required to establish threshold levels for mutation or tumor induction are prohibitive, and the debate thus becomes academic. However, for purposes of decision making and many types of relative risk estimation it is worthwhile to consider the concept of "no-effect values." These values represent concentrations of the test substance where health effects, if they exist, are less than the normal background variation in effects in the general population.

Considering the available data, certain broad generalizations can be made regarding thresholds for mutation induction.

1. Gene mutation induction at high dose levels appears to be a one-hit phenomenon that extrapolates linearly to zero dose. This conclusion is derived from experiments at concentrations where the DNA repair system is at or near saturation. Thus, the influence of repair on mutation induction is near zero. The effect of DNA repair processes at the low end of the dose range may modify the shape of the dose-response curve. In fact, data from *in vitro* assays using normal cells and repairless target cells (surrogates for high-dose repair saturation) support nonlinearity of response at low concentrations (Figure 5.5).

2. Agents which modify the mutagenic activity of chemicals (co-mutagens) generally have no-effect levels. For example, mutagenicity studies combining initiating agents and promoting agents have demonstrated modulation of mutagenic responses. These agents act through biochemical triggers that are clearly concentration dependent, and low concentrations of these modulating agents are usually not active enhancers.

3. Chemicals which are specific for clastogenic effects via disruption of spindle apparatus (e.g., colchicine and Benomyl) appear to have clear-cut thresholds.[25]

4. The approach used in interpretation of dose response data may yield different conclusions regarding threshold responses. Figure 5.6 shows two possible interpretations of a set of epidemiological data.

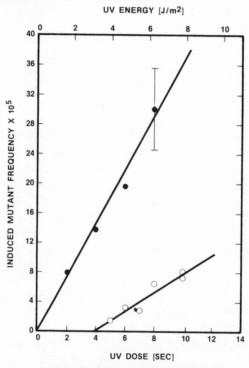

FIGURE 5.5. The effect of DNA repair on the mutation frequency in human fibroblasts. An apparent "no effect" dose is observed in normal and repair-deficient human cells. The ability of a cell to repair DNA damage clearly shifts the slope of the dose–response curve. Data courtesy of B. Myhr et al.: Ultraviolet mutagenesis of normal and xeroderma pigmentosum variant human fibroblasts. Mutat. Res. 62:341–353, 1979.

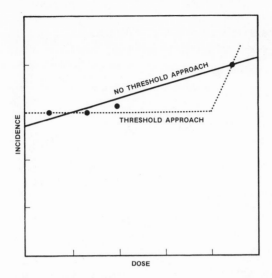

FIGURE 5.6. Two views of test data from a multidose study. The same set of data is shown graphed in two ways. Each method of data presentation supports a different interpretation of the available data.

Assuming that oncogenic transformation is dependent on a genotoxic lesion, the factors previously described for mutagenic thresholds will also apply for carcinogenic thresholds. Since it is unlikely that definitive animal studies will be conducted to prove nonlinearity of mutagenic responses, general agreement on risk-assessment models and levels of acceptable risk need to be defined by the scientific community.

REFERENCES

1. Abrahamson, S., Bender, M. A., Conger, A. D., and Wolff, S.: Uniformity of radiation-induced mutation rates among different species. *Nature (London)* **245**:460–462, 1973.
2. Ashby, J., and Styles, J. A.: Does carcinogenic potency correlate with mutagenic potency in the Ames assay? *Nature (London)* **271**:452–455, 1978.
3. Baker, T. G.: Effects of ionizing radiations on mammalian oogenesis: a model for chemical effects. *Environ. Health Perspect.*, **24**:3137, 1978.
4. Carter, C. O.: The relative contribution of mutant genes and chromosome abnormalities to genetic ill-health in man. In *Progress in Genetic Toxicology*, Vol. 2 (D. Scott, B. A. Bridges, and F. H. Sobels, eds.), Elsevier/North-Holland, Amsterdam, p. 114, 1977.
5. Clive, D., Johnson, K. O., Spector, J. F. S., Batson, A. G., and Brown, M. M. M.: Validation and characterization of the L5178Y/TK+/− mouse lymphoma mutagen assay system. *Mutat. Res.* **59**:61–108, 1979.

6. Committee for a Study on Saccharin and Food Safety Policy: Saccharin: Technical assessment of risks and benefits, Report No. 1. Assembly of Life Sciences/Institute of Medicine, National Research Council, National Academy of Sciences, Washington, D.C., November 1978.

7. Cuddihy, R. G., McClellan, R. O., and Griffith, W. C.: Variability in target organ deposition among individuals exposed to toxic substances. *Toxicol. Appl. Pharmacol.* **49**: 179–187, 1979.

8. Dean, B. J., and Senner, K. R.: Detection of chemically-induced somatic mutation in Chinese hamsters. *Mutat. Res.* **46**:403–407, 1977.

9. Dean, B. J., and Senner, K. R.: Detection of chemically-induced mutations in tissues of Chinese hamsters. In *Progress in Genetic Toxicology*, Vol. 2 (D. Scott, B. A. Bridges, and F. H. Sobels, eds.), Elsevier/North-Holland, Amsterdam, pp. 201–206, 1977.

10. de Serres, F. J.: Report of the Conference on Modification of Mutagenic and Carcinogenic Activity. *Mutat. Res.* **54**:197–202, 1978.

11. Fabro, S.: Penetration of chemicals into the oocyte, uterine fluid, and preimplantation blastocyst. *Environ. Health Perspect.* **24**:25–29, 1978.

12. Lee, I. P., and Dixon, R. L.: Factors influencing reproduction and genetic toxic effects on male gonads. *Environ. Health Perspect.* **24**:117–127, 1978.

13. Lee, W. R.: Dosimetry of chemical mutagens in eukaryote germ cells. In *Chemical Mutagens: Principles and Methods for Their Detection*, Vol. 5 (A. Hollaender and F. J. de Serres, eds.), Plenum Press, New York pp. 177–202, 1978.

14. Loveless, A.: Possible relevance of 06 alkylation of deoxyguanosine to the mutagenicity and carcinogenicity of nitrosamines and nitrosamides. *Nature (London)* **223**:206, 1969.

15. Malling, H. V., and Valcovic, L. R.: New approaches to detection of gene mutations in mammals. In *Advances in Modern Toxicology*, Vol. 4 (G. Flamm and M. Mehlman, eds.), Hemisphere Press, New York, pp. 149–171, 1978.

16. Meselson, M., and Russell, K.: Comparisons of carcinogenic and mutagenic potency. In *Origins of Human Cancer, Book C, Human Risk Assessment*, Vol. 4 (H. H. Hiatt, J. D. Watson, and J. A. Winsten, eds.), Cold Spring Harbor Laboratory, Cold Spring Harbor, N.Y., pp. 1473–1481, 1977.

17. Newcombe, H. B.: Methods of obtaining epidemiological data for mutation risk estimation. In *Handbook of Mutagenicity Test Procedures* (B. J. Kilbey, M. Legator, W. Nichols, and C. Ramel, eds.), Elsevier, Amsterdam, pp. 461–476, 1977.

18. Russell, L. B.: Numerical sex-chromosome anomalies in mammals: Their spontaneous occurrence and use in mutagenesis studies. In *Chemical Mutagens: Principles and Methods for Their Detection*, Vol. 4 (A. Hollaender, ed.), Plenum Press, New York, pp. 55–91, 1976.

19. Russell, L. B.: Validation of the *in vivo* somatic mutation method in the mouse as a prescreen for germinal point mutations. *Arch. Toxicol.* **38**:75–85, 1977.

20. Russell, L. B.: Somatic cells as indicators of germinal mutations in the mouse. *Environ. Health Perspect.* **24**:113–116, 1978.

21. Russell, W. L., Huff, S. W., and Gottlieb, D. J.: The insignificant rate of induction of specific locus mutations by five alkylating agents that produce high incidences of dominant lethality. *Biol. Div. Ann. Prog. Rep.*, ORNL-4535, pp. 122–123, 1969.

22. Sankaranarayanan, K.: Protection against genetic hazards from environmental chemical mutagens: experience with ionizing radiation. In *Progress in Genetic Toxicology*, Vol. 2 (D. Scott, B. A. Bridges, and F. H. Sobels, eds.), Elsevier/North-Holland, Amsterdam, pp. 77–93, 1977.

23. Schalet, A. P., and Sankaranarayanan, K.: Evaluation and reevaluation of genetic radiation hazards in man. I. Interspecific comparison of estimates of mutation rates. *Mutat. Res.* **35**:341–370, 1976.

24. Searle, A. G.: The specific locus test in the mouse. *Mutat. Res.* **31**:277–290, 1975.
25. Seiler, J. P.: Apparent and real thresholds: A study on two mutagens. In *Progress in Genetic Toxicology*, Vol. 2 (D. Scott, B. A. Bridges, and F. H. Sobels, eds.), Elsevier/North-Holland, Amsterdam, pp. 233–238, 1977.
26. Strauss, G. H., and Albertini, R. J.: 6-Thioguanine resistant lymphocytes in human peripheral blood. In *Progress in Genetic Toxicology*, Vol. 2 (D. Scott, B. A. Bridges, and F. H. Sobels, eds.), Elsevier/North-Holland, Amsterdam, pp. 327–334, 1977.
27. Sugimura, T.: Let's be scientific about the problem of mutagens in cooked food. *Mutat. Res.* **55**:149–152, 1978.
28. Valcovic, L. R.: Animals models used in genetic risk estimation. *Proceedings of the 9th Annual Conference on Environmental Toxicology*, U.S. Air Force Aerospace Medical Research, AMRL-TR-79-68, 1979, pp. 235–242.
29. Vogel, E.: The relation between mutational pattern and concentration by chemical mutagens in *Drosophila*. In *Screening Tests in Chemical Carcinogenesis* (H. Bartsch, R. Montesano, and L. Tomatis, eds.), IARC Scientific Publication No. 12, pp. 117–132, 1976.
30. Wiemann, H., and Lang, R.: Strategies for detecting heritable translocations in male mice by fertility testing. *Mutat. Res.* **53**:317–326, 1978.
31. Wilson, J. G.: *Environment and Birth Defects*, Academic Press, New York, 1973.
32. Würgler, F. E., Sobels, F. H., and Vogel, E.: *Drosophila* as assay system for detecting genetic changes. In *Handbook of Mutagenicity Test Procedures* (B. Kilbey, C. Ramel, M. Legator, and W. Nichols, eds.), Elsevier/North-Holland, Amsterdam, pp. 335–373, 1977.

CHAPTER 6

Applications of Genetic Toxicology to Environmental and Human Monitoring

INTRODUCTION

Occupational and environmental monitoring are two applications of genetic toxicology that are currently undergoing serious examination for mass application. Industrial emission products are continuously released into the environment at low levels over a number of years, and there is a need to evaluate their potential toxicity for individuals residing within the surrounding areas. In some cases, genetic toxicology offers the only opportunity to develop a toxicologic assessment of industrial and automotive emissions released into the environment, because the sample volumes collected are often less than those needed for whole-animal toxicology. *In vitro* assays can provide an estimate of the toxicologic potential of these emissions without the large sample size associated with *in vivo* assays. The economic, short-term test period and small sample size requirements of *in vitro* assays also make them ideal for other monitoring situations such as:

1. Quality-control techniques for batches or lots of a finished product. The goal is to detect the presence of genotoxic impurities formed during production or storage. Occasionally, changes in a chemical synthesis or manufacturing process will introduce toxic impurities into technical-grade products. Short-term genetic assays can be used to monitor batches or lots for such agents.
2. Industrial emissions change from time to time, again, depending on the process in use at any given time or the effectiveness of pollution

control devices. Thus, periodic sampling is desirable to determine chronological emission problems. Short-term tests can develop this type of data.

3. Finished commercial products are generally subjected to some type of toxicologic evaluation. The extent of the evaluation depends on the general intended use or the regulatory agencies under which the material falls. However, it is unusual to subject all intermediates used in the synthesis of the final product to toxicologic evaluation, even though workers may have occasional exposure to them. Again, short-term test results can be useful in developing a data base on a series of intermediates that can guide the development of worker exposure limits.

Occupational application of genetic tests (i.e., worker monitoring) involves different methods than those used in environmental evaluation, and also presents additional complications. Most monitoring of workers has involved analysis of chromosomes.[5] New cytologic techniques to monitor worker exposure have recently been evaluated in pilot studies with encouraging results.

MONITORING TECHNIQUES

There are two levels of monitoring which can be followed to assess the likely effects of environmental agents on workers. The first approach is to monitor the immediate environment (e.g., the occupational setting) for mutagens. The other alternative is to monitor the individuals working in a specific environment. In the first approach, the assumption is that if mutagens are discovered in the workplace, controls can be applied to the presumed source of the mutagens. In this manner, all workers will be protected to the extent that the source controls are active. This approach is also relatively inexpensive compared to worker population monitoring.

Monitoring individual workers has been proposed to compensate for individual heterogenicity. Exposure of a group of individual humans to a chemical may result in a wide range of responses depending on the lifestyle of the individuals (smokers vs. nonsmokers), genetic predisposition (high-risk individuals), and work habits (conscientious use of protective equipment). Individual monitoring for exposure using noninvasive or mildly invasive sample collecting is considerably more costly than general monitoring of the work environment.

Table 6.1 lists some of the methods available for environmental and human monitoring. Many of these methods are still developmental and are not ready for application,[1] while others such as SCE and spermhead morphology analyses are currently under examination in pilot studies. Large-

TABLE 6.1

Techniques Proposed for Environmental and Human Monitoring

Environmental monitoring	Individual human monitoring[a]	
	Sample source	Technique
Work/environmental sample collection subjected to predetermined in vivo, in vitro, and submammalian assays	Blood	Chromosome aberrations[5] Micronuclei[5,9] SCE[11] Alkylated hemoglobulin[7,13]
Particulate collection Condensation samples Solid/liquid process materials Samples can be evaluated directly or	Urine	Assessment of mutagens excreted via assays using nonhuman target cells[7,13]
following concentration, fractionation or extraction procedures	Feces	Assessment of mutagens excreted via assays using nonhuman target cells[7]
Upper limits of release can be set on the basis of available control technology. Levels which exceed these limits can trigger more intense monitoring on an individual basis.	Sperm (semen)[b]	Spermhead abnormalities[10] Fluorescent Y body technique[4]

[a] Results must be corrected for effects not related to occupation.
[b] These techniques are strictly limited to males and thus are unsatisfactory for general population screening.

scale evaluations of human populations for chromosome aberrations have been undertaken with generally disappointing results due to uncertainties in data analysis resulting from inadequate sample size.

Adjustment for nonoccupational factors is important, since data have already been collected showing that smokers have higher levels of urine mutagens,[2] SCEs,[2] and abnormal sperm[10] than do nonsmokers, and that the consumption of cooked beef increases the level of fecal mutagens[3] (H. F. Mower, personal communication). These results not only demonstrate the usefulness of the techniques for population monitoring, they also establish the need to develop good control baseline data for various human populations before the techniques are used to monitor the workplace.

Occupational Monitoring

Current industrial medicine practices encourage health effects monitoring for occupationally produced disease or toxicity. For the most part, this monitoring assesses acute effects such as live enzyme and respiratory functions and provides little information concerning exposures that may produce hereditary effects or irreversible toxicity. Several approaches have been suggested by various investigators. Most of these approaches employ cytologic analysis of cells readily obtainable from the human body. Sputum and urine cell cytology have been used as indicators of early precarcinogenic states.

These methods consider the morphology and staining characteristics of epithelial cells recovered normally in the sputum and urine. They are rapid and the samples are easily obtained, but their reliability is uncertain. This approach also fails to provide predictive information, since the atypical cells are generally a sign that the disease has already progressed to a significant degree. Since genetic end points correlate so well with other chronic diseases, genetic testing on worker populations has been initiated in some industrial settings.

Cytogenetic analysis has long been the conventional approach to human monitoring for genetic damage. Analysis of peripheral blood lymphocytes for structural or numerical aberrations is conducted as part of pre- and postemployment worker evaluation or may be used to assess chromosome damage following accidental exposure to suspect genotoxic chemicals or radiation.[5,7] Numerous limitations are associated with human lymphocyte analysis.

1. The typical variability between individuals' "background noise" in a population is relatively high using typical sample sizes, and small changes in aberrations resulting from low-level exposure are virtually impossible to detect. One method of reducing background noise is to score only for a particular type of aberration that has a predictable frequency and represents an aberration class that is clearly deleterious. Dicentric chromosome aberrations are generally used in this regard; their normal spontaneous occurrence in humans is low and relatively stable, and they are associated with the induction of translocations, a type of aberration known to be transmissible. Thus, levels only slightly increased over that of the background can be accurately detected. Obviously, because of the low spontaneous frequency, about 3000–4000 metaphase cells per individual must be scored, making this technique very time-consuming and costly.

2. Chromosome analysis is to a large extent dependent on the investigator; that is, the concordance between cytogeneticists for the classification of specific aberrations is not very great. The problem is much the same as that encountered by pathologists reviewing tissue specimens.

The results of a collaborative study involving six laboratories is shown in Figure 6.1. The compound used in this study was TEM, a potent polyfunctional alkylating agent. Even when the test material was highly clastogenic, there was substantial variability between laboratories. Much of this variability was a result of aberration interpretation. Therefore, staining techniques employed and technician training are critical to uniformity of evaluation. One of the most widely publicized incidents of a misinterpreted cytogenetic analysis involved the use of spray adhesives employed in art and craft work. The results of preliminary cytogenetic studies conducted on individuals exposed to the vapors of specific adhesive sprays were reported to

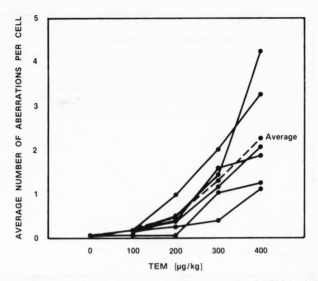

FIGURE 6.1. Comparative rodent bone marrow cytogenetic study of TEM in six laboratories. The data illustrate the typical variation among laboratories for a potent clastogenic agent. From Killian, D. J., *et al.*: A collaborative study to measure interlaboratory variation with the *in vivo* bone marrow metaphase procedure. In *Handbook of Mutagenicity Test Procedures* (B. J. Kilbey, M. Legator, W. Nichols, and C. Ramel, eds.), Elsevier Scientific Publishing Co., Amsterdam, New York, Oxford, pp. 234–260.

show significant increases in chromosome aberrations. As a result, the Consumer Product Safety Commission instituted a ban on the sale of these adhesives and ordered a recall of materials from retail stores. Subsequent review of the original slides to verify the initial reports was not successful, and upon careful analysis of the methods used by the investigator, experts agreed that the chromosome spreads were improperly interpreted.

3. Most data suggest that chromosome and chromatid breaks, chromosomal deletions, translocations, and other rearrangements result from relatively extensive damage to DNA and/or chromosomal proteins. These chromosomal lesions probably result from multiple "hits," or relatively high levels of alkylation. More subtle changes, such as those involved in gene mutation, may be increased significantly by exposure to levels of mutagens that are below the limit of detection in cytogenetic analysis, as illustrated in Figure 6.2. Thus, the detection of effects in humans exposed at low dose levels of potentially hazardous agents may go undetected if chromosome analysis is relied upon as the sole method for detecting genotoxicity.

4. Cytogenetic analysis of workers on a routine basis is labor intensive and extremely expensive compared to the typical health examinations

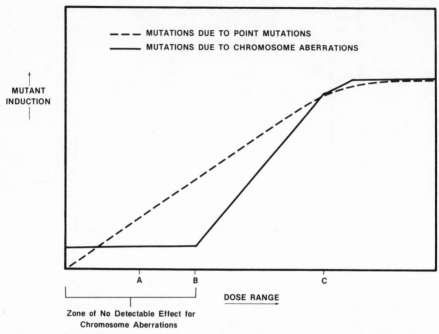

FIGURE 6.2. Hypothetical dose–response curve illustrating a situation in which chromosome analysis yields an incorrect negative designation for a mutagenic chemical. The test agent may be incorrectly declared negative if concentration "B" represents the maximum tolerated dose. If higher concentrations such as "C" were possible, the chemical would be detected in either type of test method.

performed routinely by occupational physicians. The ideal genetic assay for occupational monitoring would be rapid, inexpensive, highly objective, and predictive. One technique, SCE detection, appears to have most of these desired properties.

The SCE method using human lymphocytes recovered from test populations is currently undergoing numerous trial studies, and the results appear promising. Early studies on hospital patients receiving chemotherapeutic alkylating agents showed increased SCE.[6] Another pilot study with actual occupational exposures demonstrated a significant elevation of SCEs in a population of petroleum workers (A. Carrano, personal communication). The former results were not surprising, since many chemotherapeutic drugs are suspect human mutagens and carcinogens on the basis of *in vitro* studies.[8] The significant facts derived from these studies are that human populations are amenable to sensitive and rapid assessment with the technique. Particular features of these studies appear encouraging, however. The studies showed that (1) analysis of SCEs is rapid and not subject to the same spontaneous background variability as conventional chromosome analysis;

(2) blood samples could be collected, transported over long distances, and grown in culture under relatively uniform conditions; and (3) effects from low-level occupational exposures can be detected with a degree of sensitivity unequaled by conventional cytogenetic analysis.

Analysis of SCE in human cells may also play a role in establishing the dose received, or at least in identifying affected individuals following accidental exposure to known genotoxic substances. For example, as shown in Figure 6.3, the target site (chromosome) dose on a somatic cell basis may be

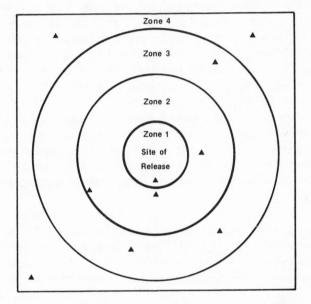

▲ = Hypothetical location of workers in relation to the site of release.

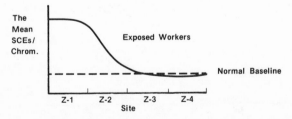

FIGURE 6.3. The use of SCE analysis to measure the chromosome dose of a mutagen received at various sites within the vicinity of an accidental release. The top illustration gives the location of the workers at the time of release, and the bottom portion shows the hypothetical profile of SCEs scored in workers from each zone. The individuals in Z-3 and Z-4 did not appear to receive an effective dose (DNA hit) of the chemical. The dashed line identifies the control range.

assessed by the degree of the increase in SCE in workers located at various distances from the primary release site. This analysis is rapid and sensitive and can be completed very shortly after the actual exposure has occurred (about 2–3 days). Based on this information, workers requiring immediate attention or long-term follow-up can be identified. There are probably no other available tests for genetic monitoring which more closely meet the requirements of a human dosimeter than does the sister chromatid exchange technique. A relatively new method measuring target cell UDS *in vivo* may be available in the near future.

All the human monitoring techniques listed in Table 6.1 should be carefully evaluated in extensive pilot study programs before being considered acceptable as occupational monitoring techniques. The two tests involving analysis of human sperm are associated with specific problems. The problems primarily consist of the uncertain availability of samples and the possibility of limited accessibility of the test agent to gametes within the gonads. Even with good pilot data, the interpretation of positive findings in workers will require intricate use of scientific and legal expertise. Current experience with health monitoring will not be useful in guiding users of this technology, since the implications of risk may be to individuals in subsequent generations. The development of monitoring techniques must be coordinated with the development of risk assessment models and societal/ethical considerations for future generations.

The findings in the petroleum workers, who were presumed to have been chronically exposed to genotoxic agents, suggest that SCEs persist for some time following chemical insult. Such data are consistent with the experimental work of Stetka and Wolff using multiply dosed rabbits.[12] In these studies, it was demonstrated that sequential injections of 3-methyl-

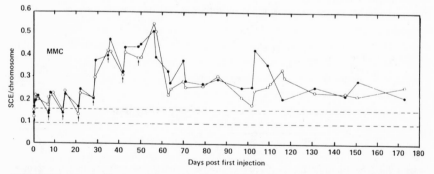

FIGURE 6.4. The production of long-lived SCEs in rabbit lymphocytes by repeated exposure to mitomycin C. Repeated exposure induces persistent increases in SCE frequency after the fourth injection. The effect may be due to persistent cells or to damage inflicted on stem cells. The dashed lines represent the normal SCE baseline and it can be observed that after termination of repeated dosing the SCE frequency does not return to that level. From D. Stetka.[11]

TABLE 6.2

Sources of Environmental Contamination and Federal Programs for Assessment

Source of environmental contamination	Program for assessment	Types of studies conducted
Administration of pesticides	EPA; Federal Insecticide, Fungicide, Rodenticide Act	1–6
Solid waste disposal	EPA; Resource Conservation Recovery Act	1, 2, & 4
Stationary site industrial emissions	EPA; Environmental Assessment Program (Level I)	1
Diesel emissions	EPA; Automotive Emissions Standards	1–6
Air and water polution	EPA; Clean Air Act.	1
	EPA; Drinking Water Act	1

[a] Studies: (1) bacteria mutagenesis; (2) mammalian cell mutagenesis; (3) *in vivo* cytogenetic evaluation; (4) primary DNA damage assays; (5) *Drosophila* SLRL assays; (6) *in vivo* germ cell assays in mammals.

cholanthrene (MCA) or mitomycin C resulted in the production of a long-lived subpopulation of SCEs (Figure 6.4). Verification of this response in human cells would mean that detection of detrimental effects induced during chronic exposure might be possible. Again, it should be emphasized that SCE may not be a detrimental lesion, but that it can be used as a signal that other potentially hazardous DNA changes are being induced by the test chemical.

Environmental Monitoring

Assessment of the genotoxicity of agents entering the environment as industrial emissions, waste products, and emissions from gasoline and diesel internal combustion engines is an ever-increasing problem. Discovery of the magnitude of negative effects of chemical dumping at sites such as Love Canal in New York has raised serious concern about our capability for handling toxic wastes, and about the effects of chronic human and ecosystem exposure to mixtures of mutagenic, teratogenic, and carcinogenic agents. Table 6.2 identifies some of the sources of environmental contamination and the Federal programs implemented to evaluate their potential for genotoxicity. Each program employs one or more tests for genotoxicity. Some, such as the pesticide program, use genetic toxicology to assess the potential for inherited effects, while others use *in vitro* tests as indicators of carcinogenic potential.

A unique feature of some of these programs is that the assessments for genotoxicity are performed on complex mixtures. Mixtures of environmental origin may contain from several to hundreds of individual agents. This added dimension generates testing problems never encountered in single-compound evaluations. Little is known, for example, about the interactions of chemicals in mixtures. Inhibitory and synergistic reactions may occur. Some components of the mixture that are toxic but not mutagenic may mask less toxic components that are mutagenic. Consideration must be given to whether or not the samples will be tested as closely as possible to the form they are in when collected, or whether samples need to undergo concentration, fractionation, or extraction procedures. Sampling techniques, as well as the transport and storage of samples prior to testing, are critical for accurate toxicity evaluation.

The general flow of a sample through the environmental assessment

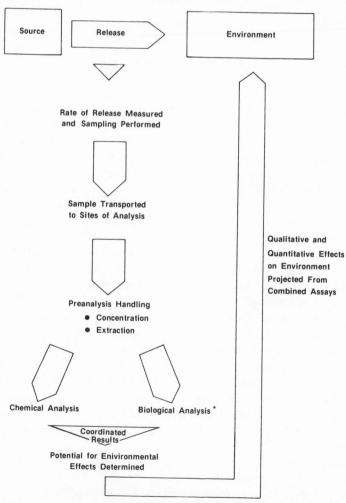

FIGURE 6.5. Typical steps involved in environmental assessment of industrial emissions. * Both human health effects and ecological effects are studied.

TABLE 6.3

Types of Tests Performed in Making Environmental Assessment Determinations

Chemical	Biological	Fate
Samples fractionated and fraction subjected to qualitative and quantitative analysis to detect known toxicants Chemical data used to supplement biological data to focus on most active toxicants; evaluate methods for removal or inactivation	Biological data developed on health and ecological effects Mutagenesis evaluations are included in health effects testing along with acute toxicity determinations Ecological testing to assess aquatic and terrestrial effects	Rate of release is measured to estimate primary exposure Persistence and bioaccumulation are measured to assess total load in the environment following release Accumulation by the food chain and water sources; used to estimate secondary exposure via recycling

program is shown in Figure 6.5. Most testing approaches for the programs in Table 6.2 contain common features, although each has unique characteristics depending on the nature of the emission source. Without applying these types of predictive genetic toxicology tests, information related to chronic irreversible toxic effects of fugitive emmissions would be impossible to accumulate.

Genetic toxicology is rapidly becoming a valuable tool in occupational and environmental assessment programs. Rapid performance time, comparatively low cost per sample tested, and predictive performance when used in a battery enable an assessment of the risks in these two areas that was previously impossible. There can be little doubt that this information will tell us a great deal about the effects of human industrialization on ourselves and our environment.

One of the most ambitious ventures listed in Table 6.2 is the EPA's Environmental Assessment Program monitored by the Industrial Environmental Research Laboratory (IERL). This program is attempting to monitor solid, liquid, and gaseous emissions from all types of industrial sites for their relative toxicity for humans and the environment. The program represents a large-scale application of genetic toxicology to environmental monitoring, and the eventual adoption of genetic testing on a routine basis will probably be derived from this effort. Table 6.3 defines how the biological and chemical information developed by this program are used.

REFERENCES

1. Beatty, R. A.: F-bodies as Y chromosome markers in mature human spermheads: A quantitative approach. *Cytogenet. Cell Genet.* **18**:33–49, 1977.

2. Bridges, B. A., Clemmesen, J., and Sugimura, T.: Cigarette smoking—does it carry a genetic risk? International Commission for Protection against Environmental Mutagens and Carcinogens Publication No. 3, *Mutat. Res.* **64**:71–81, 1979.

3. Bruce, W. R., Varghese, A. J., Furrer, R., and Land, P. C.: A mutagen in the feces of normal humans. In *Origins of Human Cancer*, Vol. 4 (H. H. Hiatt, J. D. Watson, and J. A. Winsten, eds.), Cold Spring Harbor Conferences on Cell Proliferation, Cold Spring Harbor Laboratory, Cold Spring Harbor, N. Y., pp. 1641–1646, 1977.

4. Kapp, R. W., Jr., Picciano, D. J., and Jacobson, C. B.: Y-chromosomal nondisjunction in dibromochloro-propane-exposed workmen. *Mutat. Res.* **64**:47–51, 1979.

5. Kilian, D. J., and Picciano, D.: Cytogenetic surveillance of industrial populations. In *Chemical Mutagens: Principles and Methods for Their Detection*, Vol. 4 (A. Hollaender, ed.), Plenum Press, New York, pp. 321–339, 1976.

6. Lambert, B., Ringborg, U., Harper, E., and Linblad, A.: Sister chromatid exchanges in lymphocyte cultures of patients receiving chemotherapy for malignant disorders. *Cancer Treatment Reports* **62**:1413–1419, 1978.

7. Legator, J. S., Truong, L., and Connor, T. H.: Analysis of body fluids including alkylation of macromolecules for detection of mutagenic agents. In *Chemical Mutagens: Principles and Methods for Their Detection*, Vol. 5 (A. Hollaender and F. J. de Serres, Eds.), Plenum Press, New York, pp. 1–23, 1978.

8. Matheson, D., Brusick, D., and Carrano, R.: Comparison of the relative mutagenic activity for eight antineoplastic drugs in the Ames *Salmonella*/microsome and TK+/− mouse lymphoma assays. In *Drug Chem. Toxicol.* **1**:277–304, 1978.

9. Obe, G., Beek, B., and Vaidya, V. G.: The human leukocyte test system. III. Premature chromosome condensation from chemically and X-ray-induced micronuclei. *Mutat. Res.* **27**:89–101, 1975.

10. Soares, E. R., Sheridan, W., Haseman, J. K., and Segall, M.: Increased frequencies of aberrant sperm as indicators of mutagenic damage in mice. *Mutat. Res.*, **64**:27–35, 1979.

11. Stetka, D. G.: Sister chromatid exchange in animals exposed to mutagenic carcinogens—an assay for genetic damage in humans. Presented at Dayton Aerospace Symposium, Dayton, Ohio, August 1979.

12. Stetka, D. G., and Wolff, S.: Sister chromatid exchanges as an assay for genetic damage induced by mutagen-carcinogens. II. *In vitro* test for compounds requiring metabolic activation. *Mutat. Res.* **41**:343–350, 1976.

13. Truong, L., Ward, J. B., Jr., and Legator, M. S.: Detection of alkylating agents by the analysis of amino acid residues in hemoglobin and urine. *Mutat. Res.* **54**:271–281, 1978.

CHAPTER 7

The Genetic Toxicology Laboratory

It is difficult to formulate specifications for genetic toxicology laboratories. Laboratories vary in scope and purpose. However, certain aspects should be rather universal in nature. Excluding considerations of personnel, probably the single most critical parameter is the one of safety. The seriousness of the type of toxicity resulting from exposure to highly mutagenic and carcinogenic agents must be factored into the types of facilities and equipment used to perform genetic studies.

Because of the varied nature of the assays conducted for genetic evaluation, several types of laboratory facilities are required to perform the entire range of tests. These facilities consist of separate areas for microbiology, tissue culture, cytogenetics, biochemistry, and possibly *Drosophila*. Animal facilities will also be required if *in vivo* risk assessment studies are to be conducted or microsomal activation preparations are going to be made. Support facilities for media preparation, glassware washing, and chemical storage are generally located in close proximity to the testing facility.

GENERAL LABORATORY AREAS

All laboratories should be constructed and equipped to facilitate testing under safe conditions. This means that testing areas should have limited access and that the potential hazards associated with the work should be clearly identified (Figure 7.1). Most dosing and manipulation involving chemicals should be performed under a certified Class II, Type B hood or an equivalent hood (as shown in Figure 7.2). This hood should be vented

NO   ADMITTANCE

Authorized   Personnel   Only

Cancer - Suspect   Agents

FIGURE 7.1. Warning of potential hazards. Signs such as this should be used to restrict entry into laboratory areas. Adaptations of such signs may also designate the type of safety equipment required for entry (i.e., chemical cartridge respirator required). Use of such signs is necessary to protect technicians and nontechnical staff. The level of restrictions must be clearly stated.

FIGURE 7.2. Examples of safety hoods for *in vitro* and animal exposures. The photograph on the left shows a Class II Type B hood used in cell culture *in vitro* tests. The air is not recycled but is vented through HEPA and charcoal filters. The photograph on the right shows the administration of a potentially toxic material to a small animal. This hood is also vented to protect the technician and the work area.

FIGURE 7.3. Two extremes in laboratory garments. The technician is dressed for both routine laboratory duties in nonrestricted areas (left) and in restricted areas (right). The respirator protects the individual from organic vapors. Standard operating procedures (SOPs) defining the type of laboratory garments required should be available to each technician.

through HEPA and charcoal filters to the outside. If the dosing involves animals, a hood system of the type shown in Figure 7.2 must be located near the animal holding area. Since animals may be held for 24 hr dosing before placement in the animal rooms, it is best to have the short-term holding area adjacent to the dosing hood. Much of the initial dose is eliminated in the first overnight urine and in expelled air. This reduces the carry-over of mutagens to the general holding areas.

The degree of employee safety equipment should be consistent with the degree of potential exposure and risk. The technicians shown in Figure 7.3 represent two extremes. Air change is also important; the typical air change in laboratory areas is at least 10 complete changes per hour.

CHEMICAL STORAGE AND WASTE DISPOSAL

It is advisable to designate a chemical storage area in which all types of chemicals can be safely stored. The essential equipment in this area should consist of:

1. A ventilated chemical fume hood with a balance placed inside (Figure 7.4). This hood should be equipped with an alarm that signals complete or partial loss of air flow.

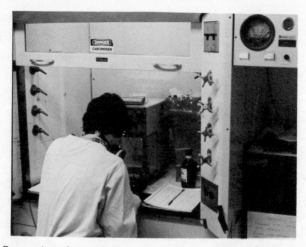

FIGURE 7.4. Preparation of test samples. The technician shown is removing and weighing a sample of test material under safe conditions. Any spill or accidental release will be contained within the chemical fume hood. The hood is equipped with an alarm that will sound if the air flow is lost.

2. A vented explosion- and fireproof cabinet for storage of chemicals under ambient conditions.
3. An explosion-proof refrigerator to store samples at 4°C.
4. An explosion-proof freezer to store samples at −20°C.

A centralized chemical storage unit minimizes technician exposure to significant quantities of pure agents in the laboratory and facilitates rapid accessibility to, and accountability of, each test sample.

Removal of residual or waste chemicals is a problem unless they can be chemically deactivated or destroyed.[1] If not, it may be necessary to contract for removal of samples. Professional hazardous-waste-removal companies prepare and dispose of toxic materials under local, state, and Federal regulations.

LABORATORY SAFETY AND EMPLOYEE MONITORING

Employees of private genetic toxicology testing laboratories probably encounter a greater number of untested chemicals than any other group of workers. It is not uncommon for two or three thousand compounds to pass through a large testing laboratory in a single year. Thus, employee safety is of critical importance.

Occupational safety should include monitoring the working environment as well as the employee. Significant levels of biologically active agents can be detected by chemical and analytical examination of laboratory-derived samples. Most of the methods employed to ensure the safety of the laboratory environment have been outlined by the Occupational Safety and Health Administration (OSHA) in "Standards for the Handling of 14 Chemical Carcinogens," and in "Guidelines for the Laboratory Use of Chemical Substances Posing a Potential Occupational Carcinogenic Risk," developed by the Laboratory Chemical Carcinogen Safety Standards Subcommittee of the U.S. Department of Health, Education and Welfare.[2] The latter document identifies specific steps in safe laboratory operations. It defines the responsibilities of each component in a laboratory hierarchy, including management, safety officer, principal investigator/study director, and technical/support staff. The principal investigator/study director is primarily responsible for implementing the laboratory safety standard operating procedures and for ensuring that all laboratory personnel follow them.

The Guidelines also recommends that preassignment medical assessments be made for all employees. This assessment is to include a work history, medical history, and physical examination. Periodic follow-up medical assessments are recommended for all employees as long as they work in the laboratory area. Surveillance of body fluids is recommended. This could very well include cytogenetic or mutagenic analyses as described in Chapter 6 (see Table 6.1). Medical records should be maintained for at least 30 years after termination.

Employee educational programs are most useful for promoting safe laboratory operations. This should include informing all employees how to find technical information about the chemicals they are handling. Laboratory personnel should also be trained in the proper use of protective clothing, laboratory equipment, and methods for personal decontamination.

Finally, the Guidelines emphasizes the need for proper labeling of potentially hazardous chemicals.[5]

The establishment of a department of occupational safety and health at most testing laboratories will ensure routine monitoring programs for all laboratory areas, major equipment, and air filtration systems, including biohazard hoods and chemical storage areas. This department should also provide emergency cleanup service in the event of laboratory spills and monitor all waste-disposal procedures.

The use of disposable laboratory garments can eliminate the problem of cleaning garments exposed to hazardous materials and reduce the spread of chemical aerosols outside the laboratory. In addition to laboratory coats, head and shoe covers and gloves (latex and neoprene types) should be provided to each technician. Each employee should be fitted with a full face-mask respirator containing cartridges to remove dust and organic vapors (Figure 7.5) for work that involves exposure to known volatile toxicants.

In addition to properly designing and monitoring the work environ-

FIGURE 7.5. Full-face respirator. Under certain conditions the potential hazard is sufficient enough that technicians must wear full-face respiratory protection. Air is filtered through the chest-pack apparatus shown in the figure. Photo courtesy of Litton Bionetics Safety Department.

ment, all filtering systems should be equipped with alarms to indicate a breakdown.

Even though chemicals with unknown toxic potential are evaluated under conditions which ensure the safety of the investigator, use of a material safety data sheet similar to the one in Figure 7.6 is advisable. This form will identify actions to be taken in emergency situations. The use of these forms is recommended even for new chemicals when all of the chemical, physical, and biological properties of the test material are not known. All employees should have access to this information.

GOOD LABORATORY PRACTICES REQUIREMENTS

Genetic toxicology studies are often conducted as part of safety evaluation profiles for chemicals that are regulated by the FDA[3] or the EPA.[4] Both of these agencies have developed and published guidelines for Good Laboratory Practices. The guidelines from the FDA became law effective June 20, 1979. Those for the EPA are expected to be binding in the near future.

The forcing factor for developing the Good Laboratory Practices was the discovery by the FDA that numerous studies submitted to that agency in support of the safety of specific products could not be verified by appropriate recorded data when the studies were audited. The development of a set of Good Laboratory Practices over the past several years has not been without controversy; even the advisability of including genetic toxicology studies under its provisions was debated. The final set of regulations, however, does include *in vitro* and other short-term testing methods.

MATERIAL SAFETY DATA SHEET

PRODUCT DESIGNATION

SECTION 1. SOURCE & NOMENCLATURE

SPONSOR'S NAME

EMERGENCY TELEPHONE NO.

ADDRESS (Number, Street, City, State, Zip Code)

CHEMICAL FAMILY

MOLECULAR WEIGHT

FORMULA

TEST ARTICLE NAME & SYNONYMS

SECTION 2. HAZARDOUS INGREDIENTS

BASIC MATERIAL	APPROXIMATE OR MAXIMUM % WT. OR VOL.	ESTABLISHED OSHA STANDARD	LD_{50}/LC_{50}		
			ORAL	PERCUT	SPECIES	CONC.

SECTION 3. PHYSICAL DATA

BOILING POINT (°F)

VAPOR PRESSURE (mm Hg)

MELTING POINT (°F)

VAPOR DENSITY (Air = 1)

SPECIFIC GRAVITY (H₂0=1) = 1)

EVAPORATION RATE (= 1)

SOLUBILITY IN WATER (Pts/100 Pts H₂0)

VOLATILE % Vol.   % Wt.

APPEARANCE AND ODOR

SOLVENTS

SECTION 4. FIRE & EXPLOSION HAZARD DATA

FLASH POINT

	UPPER
FLAMMABLE (EXPLOSIVE) LIMITS	LOWER

METHOD USED

EXTINGUISHING MEDIA

SPECIAL FIRE FIGHTING PROCEDURES

UNUSUAL FIRE AND EXPLOSION HAZARDS

PRODUCT DESIGNATION

SECTION 5. HEALTH HAZARD DATA (IF AVAILABLE)

TOXIC LEVEL	CARCINOGENIC
PRIMARY ROUTES OF ABSORPTION	SKIN AND EYE IRRITATION

RELEVANT SYMPTOMS OF EXPOSURE

EFFECTS OF CHRONIC EXPOSURE

EMERGENCY AND FIRST AID PROCEDURES

SECTION 6. REACTIVITY DATA

CONDITIONS CONTRIBUTING TO INSTABILITY

CONDITIONS CONTRIBUTING TO HAZARDOUS POLYMERIZATION

INCOMPATIBILITY (Materials to Avoid)

HAZARDOUS DECOMPOSITION PRODUCTS

SECTION 7. SPILL OR LEAK PROCEDURES

STEPS TO BE TAKEN IN CASE TEST ARTICLE IS RELEASED OR SPILLED

WASTE DISPOSAL METHOD

SECTION 8. SPECIAL PROTECTION INFORMATION

VENTILATION REQUIREMENTS	PROTECTIVE EQUIPMENT (Specify Types)
LOCAL EXHAUST	EYE
MECHANICAL (General)	GLOVES
SPECIAL	RESPIRATOR

OTHER PROTECTIVE EQUIPMENT

SECTION 9. SPECIAL PRECAUTIONS

PRECAUTIONS TO BE TAKEN IN HANDLING AND STORAGE

OTHER PRECAUTIONS

Signature   Date   Address

FIGURE 7.6. Material safety data sheet. All test substances should be accompanied by a safety data sheet describing any unusual chemical properties, special handling considerations, and recommended actions in the event of accidental human exposure.

Good Laboratory Practices defines the conditions under which a study can be considered acceptable. The fundamental principles are that a study must be conducted in accordance with a sponsor-approved study design, that there must be full documentation that each point of the study design was followed, and that the results must be reported in sufficient detail to ensure that the study was conducted in compliance with the approved study design. The typical study design called for by the Good Laboratory Practices from the FDA or the EPA includes:

1. Name and address of the sponsor.
2. Name and address of the testing facility.
3. Professional responsible for the study (principle investigator/study director).
4. Identification of the test substance, including purity, stability, and chemical and physical properties.
5. Rationale for test organism selection.
6. Dose selection and justification.
7. Description of the test system (microorganism, cell line, or animal species).
8. Protocol to be followed, including methods of data analysis.
9. Anticipated initiation and completion dates for the study.

Once the study has been initiated, all phases of data collection must be documented by the technicians performing the tests and reviewed by the supervisor or study director. Any deviations from the detailed protocol must be described, justified, and approved by the study director. An independent quality assurance unit must monitor the ongoing study to assure compliance with the Good Laboratory Practices. This unit also audits the raw data and final report to determine their conformity to the approved study design.

Once the study is complete, a report must be prepared. The final report of a typical study will include:

1. Fully documented conditions and circumstances under which the study was conducted.
2. All pertinent data, observations, and interpretations, including data from aborted trials.
3. Description of data analysis methods.
4. Amendments and adjustments to the protocol or study with justifications and signatures of responsible professionals.
5. Dated signatures of the study director and other professional personnel involved in the study.

All raw data, preserved tissue, slides, and other specimens collected during the study must be retained in archives for a specified number of years after the data have been submitted to the regulatory agency for review. Samples of the substance tested must also be maintained for the specified period.

The requirements of the Good Laboratory Practices are consistent with most good scientific investigations. The difference lies primarily with the principle that no piece of data is considered valid unless complete documentation of its origin can be provided. The development of the Good Laboratory Practices has, more than anything, stimulated the development of extensive documentation and recordkeeping (Table 7.1). This requirement for documentation of all data is beginning to be felt in other areas of research and development. Some scientific journals are beginning to request a compilation of raw data with submitted manuscripts if only graphs are shown in the report. This change will enable reviewers to better evaluate the quality of the studies being reported.

Good Laboratory Pratices will not ensure high scientific quality, but they demand sufficient quality control and documentation to help assess the quality of the study. Another feature of Good Laboratory Practices that will aid in the development of high-quality studies is the need for onsite inspection by the sponsor of a study. This forces the sponsor to interact on an intimate basis with the testing laboratory, its facilities, and the technical staff responsible for conducting the study.

The impact of Good Laboratory Practices on a growing applied science such as genetic toxicology should aid in the standardization of methodology. Still remaining is the need for standardization in data analysis and interpretations, and for perspective related to extrapolation of test results to the human situation. Extensive work on protocol standardization is also needed in this area of toxicology. As more of this type of testing becomes part of required safety evaluation for chemicals, standardization will become increasingly important to permit objective evaluations.

Presently, no formalized guidelines are available to define protocols for all types of studies employed in genetic toxicology. Most testing laboratories rely on research publications and symposia. Thus, the variation in study designs from laboratory to laboratory is likely to be substantial. The EPA Gene-Tox program described in Chapter 4 will contribute to this area by defining more-or-less standard protocols. The experience developed by a laboratory, the extent of its historical data base, and the philosophy of its staff are critical factors in study design.

There has been a general concern that with the development of standardized protocols, new test modifications and research and development will be suppressed. It is unlikely that protocol standardization will prevent basic research, but it would place data derived from a new or significantly modified assay into a less critical position than data derived from a recognized protocol. The rate at which new information is being accumulated on these tests suggests that it is too early to establish highly rigid guidelines. A minimal

TABLE 7.1

Aspects of Studies Which Must Be Verified to Meet Guidelines for Good Laboratory Practices

Personnel	Safety	Equipment	Facilities	Records	Standard operating procedures (SOP)
Sufficient number of qualified professionals to conduct studies	Proper storage and handling of chemicals and test compounds[a]	Adequate design and sufficient quantity to conduct studies	Animal areas •Caging and care in compliance with Animal Welfare Act of 1970[a]	Permanent record of signatures and initials of all employees[a]	Written procedures for •Animal care •Laboratory tests •Receipt and identification, storage, handling, administration, and disposal of test chemical
Designated study director	Hoods and airlocks available	Routinely cleaned and calibrated	•Quarantine areas available	Raw data signed and dated by technicians	•Report preparation
CVs on participants of study	Protective garments such as lab coats, gloves, shoe covers, and respirators available when required	Maintenance records must be available[a]	Waste control •Sanitary disposal of animal wastes	Records of weights, calculations, measurements, and operations or observations during study[a]	•Personnel health and safety
Quality-assurance unit must be available to review studies	No food, drink, or smoking in laboratory areas		•Safe disposal of hazardous wastes	Changes of data initialed and explained[a]	•Data handling, storage, and retrieval SOP must be available in laboratory at all times
	Chemical, physical, and toxic properties as specified in safety data forms		Adequate lab space to perform studies; Each chemical requires separate housing for animal studies	Data bound and stored at conclusion of study[a]	
			Separate administrative area	Historical file of each SOP[a]	
			Data and specimen archives	A written protocol for each study[a]	
			Analysis of feed, water, and bedding[a]	Chemical identification, purity, and stability[a]	

[a] Requires written documentation with names, dates, and explanations.

protocol could be developed and that step along with the fact that toxicology tests must be conducted under Good Laboratory Practices guidelines will at least ensure compliance with good scientific standards.

REFERENCES

1. Ehrenberg, L., and Wachtmeister, C. A.: Handling of mutagenic chemicals: Experimental safety. In *Handbook of Mutagenicity Test Procedures* (B. S. Kilbey, M. Legator, W. Nichols, and C. Ramel, eds.), Elsevier/North-Holland, New York, pp. 411–418, 1977.
2. Laboratory Chemical Carcinogen Safety Standard Subcommittee of the DHEW Committee to Coordinate Toxicology and Related Programs: Guidelines for the laboratory use of chemical substances posing a potential occupational carcinogenic risk. Revised August 1978.
3. Department of Health, Education, and Welfare, Food and Drug Administration: Good Laboratory Practices Regulations for Nonclinical Laboratory Studies. In *Fed. Regis.*, Part II, December 22, 1978.
4. Environmental Protection Agency: Proposed Health Effects Test Standards for Toxic Substances Control Act Test Rules. In *Fed. Regis.*, Part II, May 9, 1979, and Part IV, July 26, 1979.
5. Sax, N. I.: *Dangerous Properties of Industrial Materials*, 4th Edition, Van Nostrand Reinhold, New York, 1975.

Descriptions of Genetic Toxicology Assays

GENERAL CLASSIFICATION OF GENETIC TOXICOLOGY ASSAYS

A significant level of redundancy exists in the types of tests included in genetic toxicology. In an effort to group these tests into manageable assays, four major classes have been defined (Table 2.6). Two classes of tests measure genetic alterations in the form of specific locus gene mutation and chromosomal aberrations. The third category, primary DNA damage, typically measures the secondary responses produced in response to DNA lesions such as stimulation of DNA repair, somatic recombination between homologous or sister chromatids, or DNA strand breakage. The fourth category does not directly measure DNA effects at all, but assesses the ability of a test substance to transform "normal" mammalian cells into cells with neoplastic properties. It is generally assumed, however, that DNA alterations are required for transformation. This latter category of tests is used exclusively as a measure of oncogenic potential with little or no relevance to the production of heritable damage. One important feature of the oncogenic transformation series of tests is their potential for detecting agents that have promoting or cocarcinogenic properties.

DESCRIPTION OF COMMON ASSAYS FOR GENE MUTATION

Microbial

There are large numbers of point mutation assays in bacteria, yeast, and mold. Each type of organism has its own specific attributes and limitations. Mutation assays within the three phylogenetic categories of micro-organisms have been reviewed in detail.[5,6] From the array available, the following three assays can be singled out as having the essential criteria needed for genetic toxicologic studies. All have been subjected to a diverse group of chemical mutagens, they are generally reproducible, detailed protocols for each have been published, and all can be used with an S9 activation system.

   1. *Bacteria*. Ames *Salmonella*/microsome assay for reverse mutation measures mutation in a series of histidine-requiring auxotrophs of *S. typhimurium*. Mutant strains represent "hot spots" and detect both base-pair substitution and frameshift mutagens. The basic series of strains routinely employed has undergone several modifications to increase their sensitivity. The origin and development of these strains is given in Table 8.1. Figure 8.1 illustrates the enhancing effect that some of the modifications have on the sensitivity of one set of mutant strains. Most other point mutation assays that measure either forward or reverse mutation in bacteria duplicate the detection pattern of the Ames assay and can be expected to add little to the results obtained in the *Salmonella*.[19] The *E. coli* WP$_2$ series of bacteria have been used in many laboratories for chemical screening.[6] However, until recently they have not had the necessary modifications such as introduction of the pKM101 plasmid. More recent modifications in the *E. coli* strains may increase their usefulness as screening tools.

   2. *Yeast*. Both forward and reverse mutations can be detected by yeast assays. Forward mutation to canavanine resistance is a convenient and

TABLE 8.1
Genotype of the TA Strains Used for Testing Mutagens[a,b]

Additional mutations in		Contains plasmid	Histidine mutation in strain			
LPS	Repair		*hisG46*	*hisC207*	*hisC3076*	*hisD3052*
+	+	—	*hisG46*	*hisC207*	*hisC3076*	*hisD3052*
+	Δ*uvrB*	—	TA-1950	TA-1951	TA-1952	TA-1534
Δ*gal*	Δ*uvrB*	—	TA-1530	TA-1531	TA-1532	TA-1964
rfa	Δ*uvrB*	—	TA-1535	TA-1536	TA-1537	TA-1538
rfa	+	—	TA-1975	TA-1976	TA-1977	TA-1978
rfa	Δ*uvrB*	pKM101	TA-100	—	TA-2637	TA-98
rfa	+	pKM101				TA-94
+	Δ*uvrB*	pKM101	TA-2410	—	—	
+	+	pKM101	TA-92	—	—	TA-2420

[a] From Ames *et al.*[2]

[b] All strains were originally derived from *S. typhimurium* LT-2. Wild-type genes are indicated by +. The deletion (Δ) through *uvrB* also includes the nitrate reductase (*chl*) and biotin (*bio*) genes. The Δ*gal* strains (and the *rfa*, *uvrB* strains) have a single deletion through *gal chl bio uvrB*. Strains containing pKM101 carry an extra chromosomal element (plasmid) that enhances error-prone repair.

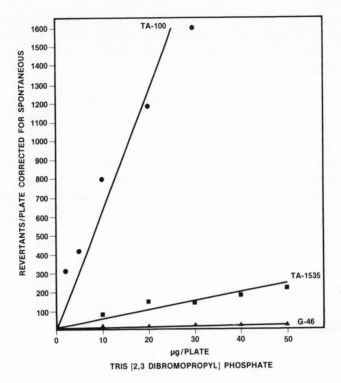

FIGURE 8.1. Differential strain sensitivity to Tris (2,3-dibromopropyl) phosphate using aroclor 1254-induced rat liver S-9 fraction. From Brusick, D. J.: In *Carcinogens, Identification, and Mechanisms of Action* (A. C. Griffin and C. R. Shaw, eds.), Raven Press, New York, pp. 93–105, 1979.

reliable test.[7] Canavanine is a toxic analog of arginine, and when placed in growth medium of normal haploid yeast [canavanine-sensitive (*can*[s])] uptake of canavanine by the arginine transport kills the *can*[s] cells. Cells which have undergone a mutation in the transport gene, rendering it inactive, synthesize arginine *de novo* and are resistant to the canavanine (*can*[r]) in growth medium. This method detects both base-pair substitution and frameshift mutagens.

Yeast assays have a general limitation in that the cell membrane restricts the permeability of certain molecules which might have genotoxic properties but cannot reach the target DNA. Numerous studies have reported significant microsomal metabolism by growing yeast cultures.[9] This might be considered either an advantage or disadvantage, since activation of chemicals is often required; however, target cell metabolism is generally not desirable, as there is a chance that it will form metabolites not produced by animal microsomal enzymes. Activation studies with yeast can be performed, but require long exposure periods to the chemical and S9 mix.

3. *Mold.* The most useful gene mutation assay in mold is the *ade3* system in *N. crassa.* This assay can provide a wealth of information regarding the mutagenic mechanisms of mutagens and can detect base–pair substitution and frameshift mutagens. Detailed protocols have been published and several modified (repair-defective) substrains have been isolated that show enhanced sensitivity.[5] Both forward and reverse mutation studies can be conducted at the *ade3* gene, and techniques for detecting multilocus effects (small internal chromosomal deletions) have also been incorporated in the assay. As in the yeast, there are problems associated with chemical permeability and endogenous target cell metabolism. Another problem involves the facilities and equipment required to perform forward mutation studies in these organisms. Most laboratories cannot conduct the test unless equipped with glassware and equipment unique for this procedure; therefore, it is presently limited to a very small number of laboratories.

Mammalian Cells *in Vitro*

Table 8.2 lists several of the common gene mutation tests conducted in cultured mammalian cells. Most of these tests utilize one of three loci. The biochemical steps associated with the use of each are given in Figure 8.2. The HGPRT and TK loci should be capable of detecting both base-pair substitution and frameshift mutagens, whereas the ouabain locus will probably be suboptimal for detecting frameshift mutagens. Frameshift mutagens often result in complete loss of enzymatic function (Figure 2.2), which in the case of ouabain would be lethal rather than mutagenic.

TABLE 8.2
Some Commonly Used *in Vitro* Mammalian Cell Lines for Gene Mutation Studies

Cell line	Genes used for mutant selection
Mouse lymphoma L5178Y	*TK*
	HGPRT
	Ouabain resistance (*OR*)
CHO	*HGPRT*
	OR
V79 hamster cells	*HGPRT*
	OR
Human lymphoblast	*HGPRT*
	OR

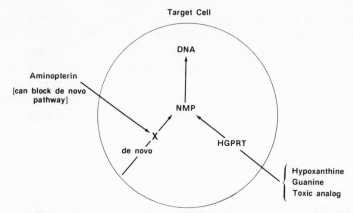

FIGURE 8.2a. Selection of mutagen-induced HGPRT phenotype. A mutation in the gene makes the target cell insensitive to the toxic selective agents, and DNA is synthesized *de novo*. Azaguanine and thioguanine are two analogs used to select for HGPRT mutants. NMP, Nucleoside monophosphate.

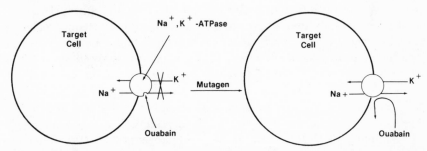

FIGURE 8.2b. Selection of mutagen-induced ouabain-resistant (Oua®) phenotypes. Ouabain affects the Na^{2+} and K^{2+} transport enzyme system. Mutant cells have an altered protein which is no longer insensitive to ouabain yet permits normal Na^{2+} and K^{2+} transport. Complete (nonleaky) mutants will probably be lethal and not detected in this assay.

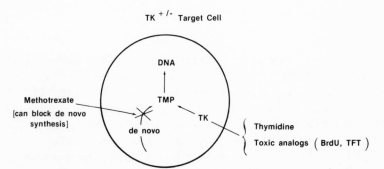

FIGURE 8.2c. Selection of mutagen-induced $TK-/-$ phenotypes. Mutant $TK-/-$ cells are resistant to the pyrimidine analogs BrdU or trifluorothymidine (TFT) and synthesize their DNA *de novo*. Nonmutant cells ($TK+/-$) will die in the presence of BrdU or TFT.

All of the techniques listed in Table 8.2 can be performed with mammalian S9 preparations similar to those used in the Ames *Salmonella* assay. The forward mutation assay using L5178Y mouse lymphoma cells is a sound test presently conducted in several laboratories, and it is being used in validation studies sponsored by the National Cancer Institute. Gene mutation induction in CHO and V79 hamster cells are included in addition to the L5178Y mouse lymphoma cells under the Gene–Tox review program (Table 4.2).

Insects

The only validated gene mutation assay in insects is the SLRL assay in *D. melanogaster*. The utilization of this technique in genetic toxicology has been the subject of more than one review.[30,31,34.] All classes of point mutations can be detected in germ cells with the *Drosophila* test, and this test has been suggested for risk estimation for heritable effects. Its major limitation is associated with the difficulties often encountered in quantitatively determining the dose actually administered to the flies by the typical feeding route. As a result, extensive preliminary toxicity and uptake studies are required prior to testing to ensure that the chemical has been taken up by the flies.

While the basic methodology of the SLRL is rather standardized, *Drosophila* strains vary from laboratory to laboratory. Recently, repair-deficient strains have been isolated, and are under analysis to determine if they will have advantages over the traditional strains.

Sampling of different stages of spermatogenesis in *Drosophila* has demonstrated germ cell stage specificity in response to mutagens, and the type of microsomal metabolism and detoxification found in both mice and flies is sufficiently close to suggest to some that *D. melanogaster* be given the status of "Honorary Mammal."

Mammals

Figure 8.3 outlines the basic study designs for the mouse specific locus assay and the mouse recessive spot test. Both assays detect morphologic changes as the result of presumed gene mutations at one of four or five gene loci.[17] In the case of the specific locus assay, the mutation is expressed as a mutant phenotype in the F_1 generation. In the spot test, coat color spots are observed on F_1 pups treated *in utero* (Figure 8.4).

The number of chemicals that can be evaluated in these tests is limited because of the need for specific inbred mouse strains and large animal facilities. The procedures, with discussions of the limitations, have been reviewed by Ehling.[15]

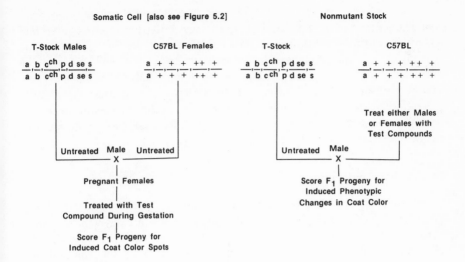

FIGURE 8.3. Comparison of the procedures for the recessive somatic spot test and the specific-locus germ-cell test. a, Aguti; b, brown; c^ch, chinchilla; p, pink eye; d, dilute; se, short ear; s, sepia; +, normal gene.

TESTS FOR CHROMOSOME ABERRATIONS

Microbial

Most microbial assays employ haploid indicator organisms and are not used for chromosome analysis. Chromosome breakage in these cells does not produce mutations but does result in lethality. Some fungi, however, are capable of measuring specific types of alterations involving chromosomes. As described in Chapters 3 and 4, *N. crassa* can detect multilocus chro-

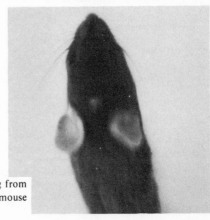

FIGURE 8.4. Coat-color spot. A coat-color spot resulting from a presumed somatic mutation is seen on the head of the mouse shown in the photograph.

mosome deletions in the *ade3* heterokaryotic system,[5] and chromosomal nondisjunction can be measured in *A. nidulans*.[3] Neither of these microbial assays is routinely employed in genetic toxicologic studies because of the organisms' distant phylogenetic relationship to animal systems and the requirements for highly specialized equipment and facilities. When considering chromosomal modifications, either structural or numeric, cell cycle mechanisms and chromosome movement are important. These processes appear to be different between fungi and animals, making extrapolation of fungal results to mammals difficult.

Mammalian Cells *in Vitro*

Almost any cell line with a well-defined karyotype could be used for *in vitro* cytogenetic analysis. Primary considerations are growth potential, chromosome number, and chromosome diversity. Rat chromosomes, for example, are almost entirely acrocentric, somewhat increasing the difficulty in scoring. CHO cells have a low number of chromosomes which are diverse in appearance, making scoring less tedious; they are therefore preferred by many laboratories for routine analyses. Studies with cell lines can be conducted with S9 microsome activation systems. Cultured human lymphocytes are often used because of their perceived relevance. Table 8.3 lists several of the cell sources commonly employed in *in vitro* cytogenetic assays. It should be stressed that *in vitro* methods for cytogenetic analysis can be misused. High concentrations of nontoxic materials may change pH or osmotic conditions to the extent that chromosome integrity is disrupted. Results from such tests are not interpretable.

TABLE 8.3
Commonly used Cells for *in Vitro* Cytogenetic Analysis

Cell (species)	Diploid chromosome number[a]	Karyotype[b]	Unique features
CHO	22 (18–22)	Abnormal	Several different chromosomes
V79 (Chinese hamster lung)	22 (21–22)	Abnormal	
L5178Y (mouse lymphoid)	40 (38–40)	Abnormal	One large metacentric chromosome
WI-38 (human)	46	Normal	
Human lymphocytes	46	Normal	

[a] Numbers in parentheses represent typical range for chromosome number in the cell line.
[b] Indicates if there are district morphological or numerical differences between the line and the cells as found in the animal of origin.

Insect Tests for Chromosome Effects

D. melanogaster is a versatile test organism which can be used to detect heritable chromosome effects, including sex chromosome (X or Y) loss and heritable translocations. Again, both effects are measured in germ cells exposed to the test substance.

The test for loss of X and Y chromosomes in *D. melanogaster* is a rapid, one-generation screen for chemicals that produce chromosome breakage or nondisjunction. In *Drosophila*, the sex of the individual is determined by the ratio of X to Y chromosomes. The possible viable combinations resulting from normal and abnormal segregation are:

Configuration	Expression
XY	Normal male
XX	Normal female
XO	Sterile male
XXY	Superfemale (abnormal but functionally unaffected)

The chromosomally unbalanced individuals, XO and XXY, are relatively unaffected because the X and Y are structurally very unlike one another; the X contains a great deal of information and the Y very little. The X is estimated to contain 20% of the total genome of the fly, while the Y contains presumably very little other than male fertility factors. Only the configurations Y and XXX are inviable. By marking the sex chromosomes with genes for readily distinguishable phenotypes, it is easy to track the progress of the treated chromosomes as they are passed to the offspring.

Reciprocal, balanced translocations can be readily measured in *Drosophila* by "engineering" stocks that have a visible mutant on their major autosomes (nonsex chromosomes). The simplest markers to use are those for eye color, since the technician can then look at only one localized body region when scoring for mutants. The genetic principle or rationale for the test is that each autosome has a homolog or "mate" that provides a similar amount and type of information. The information provided may of course be slightly different at each locus, but for the individual to survive, large portions of the chromosome can neither be deficient nor occur in excess. Two or more breaks induced in a chromosome that then rejoin improperly, transferring parts of chromosomes, will result in viable offspring only if there is an equal or reciprocal exchange of information (a translocation).

The scheme described in Chapter 9 measures the frequency of reciprocal translocations induced in wild-type males exposed to chemicals. By crossing the treated male to a female homozygous for visible mutants on the second and third chromosomes, backcrossing the sons of this cross to a female identical to the mother, and examining the F_2 progeny for the presence of flies with four eye colors (red, white, orange, and brown), one is able to determine whether the sons of the treated male carried a reciprocal translocation. Two of the eye-color classes resulting from such a male (orange and brown eyes) indicate an unbalanced amount of information. A diagrammatic representation of the crosses is provided in the protocol in Chapter 9.

Both of these end points measure chromosome alterations analogous to sex chromosome and translocation events in mice. Thus, if one assumes sufficient similarity in gametogenesis and metabolism, the *Drosophila* system may function as a model of the whole animal in measuring for heritable chromosomal alterations. The available data from testing suggests that similar results can be achieved in both test organisms.

Mammals

Rodent have long been utilized in genetic toxicology for evaluating chromosomal alterations. Most of the work has been conducted in mice, with fewer studies using rats, hamsters, and assorted other small rodents. The *in vivo* bone marrow analysis for chromosomal effects and the dominant lethal test have respectable data bases in both mice and rats. Heritable translocation analyses and sex chromosome loss are routinely conducted in mice.

Dominant Lethal Assays

Tests for dominant lethality have been conducted since 1953. They typically provide information on the fate of the zygotes produced from the sperm of treated male animals and the eggs of untreated females. The test measures death that occurs between fertilization and parturition. Events such as failure of the zygote to implant in the uterine wall or to survive to midpregnancy after implantation are detected upon necropsy of the mated females. It is generally assumed that the vast majority of such dominant lethal effects are the result of chromosome alterations, either structural or numeric. Cytogenetic analysis of preimplanted zygotes for chromosomal integrity can be performed to confirm this mechanism. This technique is extremely tedious and is not routinely performed, since it requires recovery of fertilized eggs from the fallopian tubes and uteri. Studies of this type have clearly established that structural and, especially, numerical chromosome alterations are associated with dominant lethality.

There are three rather standardized approaches to an evaluation for dominant lethality. The primary differences are shown in Table 8.4. Either

TABLE 8.4

The Major Differences between the Standard Dominant Lethal Protocol, the Extended
Dosing Protocol, and the European Work Group Protocol[16]

	Standard	Extended dosing	European
Animals per dose level[a]	10	10	50
Dosing	5 days	8–10 weeks	5 days
Mating periods (number × duration)	7 × 5 days	2 × 5 days	13 × 4 days
Rest between mating periods	2 days	2 days	—
Females per male during mating	2	2	1

[a] Better data can be obtained if proven breeder males are used.

the mouse or rat is an acceptable species; selection of species is often made to coincide with other toxicologic data.

The traditional approach to an evaluation for dominant lethality consists of an acute or subchronic exposure followed by sequential weekly matings for 7–10 weeks. Each week, the dosed males are caged with two virgin females. The females are sacrificed at midpregnancy and necropsied. By 7 weeks postexposure, all cell stages that were present at the time of exposure to the test material have traversed the entire meiotic cycle and been used to fertilize the females. If the test material has a specific effect on one or more cell stages, the effect would be observed as an elevation in dominant lethal parameters at the time of most susceptibility (Figure 8.5).

The second approach follows a scheme where the males are dosed for 7–10 weeks and then mated twice. This extended dosing method permits all cell stages to have continuous exposure spermatogenesis. All damage will accumulate in the sperm and presumably be expressed in the 2 matings. The advantages associated with this method are related to the fewer numbers of animals involved in conducting 2 matings rather than 7 or 10. The accumulation of dominant lethality in the 2 matings may produce an enhanced response if the test substance is genotoxic.

An expanded procedure has been recommended, especially by European investigators.[16] This study design increases the number of dosed males to 50 per dose group. The dosed males are subjected to sequential mating; however, the matings are made 1 : 1, male to female, and carried out on 13 consecutive 4-day cagings. The 4-day cagings are based on the estrous cycle of the mouse and the assumption that if mating occurs, it will occur during the first 3 days of caging male and female mice. This procedure is considerably more extensive than the two previously described but provides considerably more information. Using this approach, small increases over the spontaneous level can be detected with a high degree of confidence. This

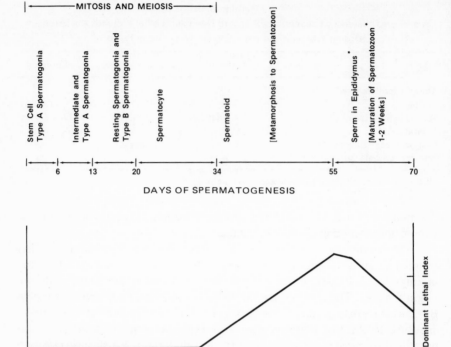

FIGURE 8.5. Stages of sperm development in the rat and its effect on the observation of dominant lethality. The dominant lethal index (dead implants/total implants) will increase at the germ cell stage (*) most sensitive to the mutagenic agent. In the above example, the mature sperm were most sensitive and an effect is seen in the first 3 weeks of mating posttreatment.

version of the test, however, requires 3500 animals when positive and negative controls and three concentrations of the test material are included in the study design.

Dominant lethal assays can be performed on both male and female mammals, but studies in females are not performed routinely because oocytes are generally less sensitive to chemicals than are sperm. It is difficult, therefore, to unequivocally discriminate between cytotoxic and true genetic effects.

The high spontaneous background of dominant lethality in rodents (6–7%) and the inability to verify heritability makes the dominant lethal assay a suboptimal method for sensitive estimations of germ cell risk unless large numbers of animals are used. Since almost all dominant lethal effects are due to chromosomal alterations, it appears that *in vivo* cytogenetic analysis (bone marrow cells) would be a more sensitive screen for clastogenic agents, thus limiting the need for dominant lethal assays for compounds which fail to show somatic clastogentic activity (see Protocol 13 in Chapter 9 for a general description).

Heritable Translocation Assay

This assay measures chromosomal translocations produced in germinal tissue which are capable of being transmitted to the next (F_1) generation. Translocation heterozygotes are characterized by reduced fertility (partial or complete); detection of the heterozygotes is made on the basis of litter size (living and dead embryos) after mating the dosed males with normal females. Confirmation is generally assessed by cytogenetic analysis of gonadal cells (spermatocyte evaluation) for recognizable meiotic chromosome figures indicative of translocations. Although direct analysis for translocation-bearing spermatocytes in F_1 progeny can be performed without the preliminary mating sequences, preselection of presumptive translocation carriers via fertility testing is the standard approach, since it reduces the number of animals that have to be cytologically analyzed.

The test uses the mouse as the target species because of the ability to clearly detect chromosome translocation figures in spermatocytes of this species, and because of cost. A typical translocation experiment may include anywhere from 4000 to 10,000 animals, depending on the level of sensitivity desired.

There is a certain analogy between the dominant lethal and heritable translocation assays, since both tests measure chromosomal damage. Only specific types of chromosomal damage, such as reciprocal translocations, are compatible with cell viability and will be transmitted to the F_1 generation. This is expressed by the very low (0.1%) frequency of translocation heterozygotes found in control populations of mice.[18]

A general outline of a mouse heritable translocation test is shown in Figure 8.6. A test requires between 30 and 35 weeks, depending upon the number of matings to be conducted. Table 8.5 is useful in developing protocols, since it defines the likelihood of detecting a given translocation rate, α, for a given population of F_1 males sampled.[18] For example, 300 F_1 males are necessary to detect an induced rate of 1–2% with a level of confidence near 95%. This depends somewhat on the criteria used to define F_1 males as fertile. A good analysis of the effect of criteria on the number of animals

WEEK	1	2	3	4	5	6	7	8	9	10	11	12	13	14	15	16	17	18	19	20	21	22
Function	Quarantine			Dose Compound						Mate F_0	Gestation and Delivery					Wean F_1 Males				Select Groups of F_1 Males		

WEEK	23	24	25	26	27	28	29	30	31		32	33	34	35	36	37	38	39	40	41	42	43	44
Function	Mate F_1	Gestation and Kill		Submit Data	Review of Data						Cytogenetic Analysis of Presumptive Translocation Heterozygotes												Final Report

← Review Data for Selection of
Presumptive Carriers

← Kill Animals/Prepare Slides

FIGURE 8.6. Typical performance schedule for the heritable translocation assay.

TABLE 8.5
Analysis of Heritable Translocation Assay: Definition of the
Sensitivity of the Test[a]

Number of F_1 males mated	Induced (true) translocation rate (%)	p value[b]	Level of significance (α)
100	1.0	0.26	$\alpha \leq 0.01$
	1.5	0.44	$C \geq 2^c$
	2.0	0.60	
300	1.0	0.80	$\alpha = 0.03$
	1.5	0.94	$C \geq 2$
	2.0	0.98	
600	1.0	0.94	$\alpha = 0.02$
	1.5	0.99	$C \geq 3$
	2.0	>0.99	

[a] From Generoso et al.[18] This table defines the likelihood of detecting a given rate at α for a given population of F_1 males. Thus, if one is looking for induced rates of 1–2%, a group of 100 animals is suboptimal; a level of 300 F_1 males gives a much more accurate assessment.
[b] Assuming the spontaneous translocation carrier frequency is 9.1×10^{-4} in untreated F_1 males, this gives the chances of detecting rates given in column 2.
[c] Minimum number of translocation carriers observed to conclude that the measured rate is significantly greater than the spontaneous rate with a probability of α.

required for heritable translocation evaluations in mice is given by Wiemann and Lang.[32]

Both Generoso et al.[18] and Weimann and Lang[32] reported that translocation carriers induced at a rate of 1–2% can be detected with a relatively high degree of confidence using reasonable numbers of animals. For example, a test with a negative control group and a single-dose level might use about 5000 animals. Dose–response evaluation, however, would involve significant time and expense.

Sex Chromosome Loss

The test for sex chromosome loss in mice is not commonly conducted in genetic toxicology laboratories. Like the *Drosophila* assay, it is based upon phenotypic genetic markers located on the X chromosome. If either sex incurs an X loss, the F_1 female progeny resultant from the loss can be identified (Figure 8.7). Sex chromosome nondisjunction is presumably detected with this approach. Nondisjunction accounts for the greatest proportion of human disorders associated with chromosomal alterations, but there is presently little evidence in mammals for chemical-induced nondisjunction.

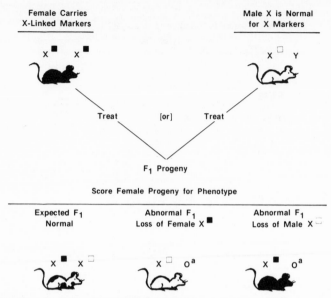

FIGURE 8.7. Sex chromosome loss in mice. This assay is based on the fact that XO female mice are viable. Distinct genetic traits placed on the two chromosomes are used to show which X chromosome is lost. The normal XX female shows both traits.

TESTS FOR PRIMARY DNA DAMAGE

This category contains a heterogeneous group of tests. Some of the tests, such as those detecting somatic recombination and DNA strand breakage, measure direct genotoxic effects. Others, such as the tests for UDS and SCE, are indirect indicators of DNA damage that may result in genotoxicity. In other words, the process of UDS is not normally considered a mechanism leading to genetic damage; rather it is the expression of the repair of presumed DNA lesions that, if they were not repaired, might lead to gene mutation or chromosomal breakage. Likewise for SCE, the process of reciprocal exchange between identical sister chromatids should theoretically not lead to genetic damage; however, some investigators have observed a general relationship between SCE induction and gene mutation induction kinetics in the same target organism.[10]

The application of primary DNA damage tests to genetic testing is currently the subject of debate. There is a general consideration that gene mutation and chromosome aberration tests should be the only ones relied upon for the assessment of genotoxic agents. Tests for primary DNA damage, although contributing to positive findings in gene mutation or chromosome aberration assays, should not stand alone as indications of potential human health hazard. In reality, it would be difficult to presume significant genetic hazard from a chemical which does not induce point mutations or chromosome aberrations but does induce SCE. This type of rationale is expressed in Table 4.9, which presents weighted values for previous primary DNA damage assays in a model scoring system.

Similar to other end points, those encompassed by the general category of primary DNA damage are detected in a variety of submammalian, *in vitro*, and mammalian species.

Microbial Tests

Bacteria

Several genera of bacteria have been employed as indicators of primary DNA damage. The basis for microbial tests rests on the ability to derive repair-deficient (usually excision) substrains from normal repair-competent strains. The two bacteria substrains should be identical in all respects except for their ability to repair DNA lesions. If the two substrains are exposed to a genotoxic material, the nonrepaired lesions in the deficient strain will result in lethality, whereas the same concentration of the genotoxic material in the normal strain is less toxic. Nongenotoxic chemicals should exhibit roughly equivalent lethality.

The effect of differential lethality is not a measure of hereditary changes, but only identifies agents which impact on DNA resulting in lethal damage. The damage may or may not be of a type thar produces hereditary effects.

Several bacteria systems, including those using *E. coli*,[26] *S. typhimurium*,[2] and *B. subtilis*,[20] are routinely employed in these tests; and interpretation of effects is based on the ratio of lethality in the two strains.

Yeast

The processes of somatic recombination are included in the group of tests which detect primary DNA damage. Somatic recombination consists of reciprocal, mitotic crossover, and nonreciprocal mitotic gene conversion assays. These tests measure DNA exchanges between segments of homologous chromatids. In essence, the processes produce homozygous states from heterozygous states in somatic cells. This process may be a factor in the expression of deleterious mutations for cancer or terata induction that are in a heterozygous state $(+/-)$ in somatic cells.[35]

Cultured Mammalian Cell Repair Assays

Several types of tests for primary DNA damage have been developed using mammalian cells. The tests can be categorized into three basic methods:

Description of test	Typical cell type employed	Method for primary DNA damage detection
UDS	WI-38 or primary rat hepatocytes	Autoradiographic analysis of [³H]thymidine incorporation into non-S-phase nuclei of treated cells
Alkaline elution technique	Any cell type	Detection of a reduction in DNA molecular size by elution of treated cells with alkaline on filters permitting small DNA species to pass through
DNA strand breakage	Any cell type	Detection of broken DNA strands by changes in DNA sedimentation in $CsCl_2$ gradients following treatment of cells with the test agent

The latter two tests are similar in nature in that each measures the ability of the test chemical to break the DNA molecule in the chromosome, producing numerous smaller-molecular-weight pieces distinguishable from the normal high-molecular-weight intact molecule. This is not analogous to chromosome breakage, since the effect is not cytologically visible and the damage is likely to be repaired. Neither of these tests is widely employed in routine screening; they are time-consuming and relatively susceptible to misinterpretation unless performed under stringent conditions by highly experienced investigators.

The DNA repair assay used by many genetic toxicology laboratories is the UDS technique. The rationale for this test is if cells are prevented from replicating (scheduled DNA synthesis) and simultaneously exposed to the test material and ³H-labeled thymidine, no significant incorporation of labeled tritium will occur unless the test substance damages DNA and thus stimulates the repair system (UDS). The stimulation of repair and subsequent incorporation of [³H]thymidine can be detected either biochemically via DNA extraction and scintillation analysis, or by autoradiographic techniques. Figure 8.8 shows the typical labeling observed in nonreplicating nuclei and nuclei in UDS.

An adaptation of the UDS autoradiographic methodology to rat primary hepatocytes has been developed by Williams.[33] This method offers two distinct advantages over the traditional UDS assay employing WI-38 cells. In the technique developed by Williams, there is no need to use chemical blockage of scheduled DNA synthesis, since: (1) the hepatocytes are nonreplicating over at least 48 hr after collection; and (2) the primary

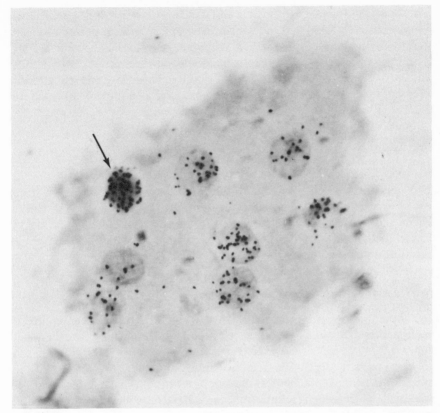

FIGURE 8.8. Grain counts in the nuclei of treated cells. The photograph shows several cells with increased radiolabel incorporated. This incorporation was the result of stimulation of the DNA repair system following compound-induced DNA damage.

hepatocytes carry out all microsomal enzyme reactions, ensuring reliable detection of metabolic reaction products. This approach appears initially, at least, to be the most reliable of the primary DNA damage assays in mammalian cells.

SCE Analysis

SCE analysis is a rapid, objective method of observing reciprocal exchanges between sister chromatids. Although the concept of SCE was established by Taylor in 1958,[29] recent staining techniques have greatly facilitated performance of this assay. The general method relies upon the phenomenon of BrdU incorporation into DNA in place of thymidine. After two rounds of cell division, the two chromatids are asymmetrically labeled with BrdU and consequently stain differently with the Hoechst stain. SCE are visualized as light–dark exchanges between chromatids stained after two mitotic divisions (see Figure 2.17). While not resulting in a genetic change, SCE appear to respond to genotoxic compounds in a dose-related fashion and can offer clear evidence of DNA exposure. Thus, SCE induction can be used both as a potential genotoxic end point and as a monitor of exposure. SCE analysis can be performed in most higher eukaryotic organisms.

Insects

As discussed earlier in this chapter, repair-deficient strains of *Drosophila* have been isolated and studied. Using the repair-sufficient strain and its repair-deficient mutant strain, comparative lethality studies can be conducted that are similar in rationale to the bacterial DNA repair assays. These *Drosophila* assays are in a very preliminary stage of development at this time and are not currently included in genetic testing programs.

In Vivo Mammalian Primary DNA Damage

Attempts to detect primary DNA damage *in vivo* have been limited to pilot studies. The alkaline elution technique was applied, by Swenberg and Petzold,[28] to DNA carefully extracted from cells of various organisms in mice exposed to mutagenic agents. The methodology is amenable to this type of *in vivo* analysis, but test conditions must be very carefully controlled since cellular damage resulting from the cell/DNA collecting methods used are likely to introduce artifacts. Dose selection must be carefully controlled as well, since cytotoxicity will result in cellular necrosis and probable DNA breakage. These effects are independent of DNA-directed damage, and lead to incorrect assumption of genetic activity. It is unlikely that this technique can be successfully applied to routine assessments of genotoxicity because of the number of variables capable of introducing artifacts.

SCE can be detected *in vivo* as well as *in vitro*. The critical aspect of *in vivo* studies is the necessity of having the target cells undergo two rounds of DNA replication in the presence of BrdU. Application of BrdU *in vivo* has been accomplished by intravenous infusion or by subcutaneous implant of BrdU tablets.[1] The latter method is the simplest and only requires minor surgery. Once the tablet is implanted, the BrdU difuses to the replicating target cells and becomes incorporated. The animal is killed at a specified time, and the target cells removed and stained. Visualization and scoring of SCE induction is similar to the methods used *in vitro*. Highly proliferative cells such as bone marrow appear to be a good source of target cells.

Additional methods have been developed to collect cells from exposed animals and perform the staining techniques by culturing the cells in

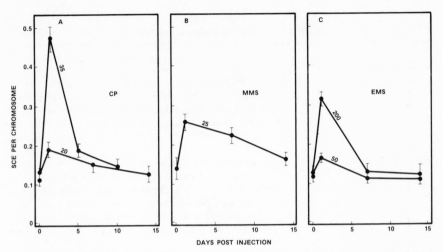

FIGURE 8.9. *In vivo* SCE induction in rabbit peripheral lymphocytes. Results from three mutagens are shown: (A) cyclophosphamide (CP) at 20 and 35 mg/kg; (B) methyl methanesulfonate (MMS) at 25 mg/kg; and (C) ethyl methanesulfonate (EMS) at 50 and 200 mg/kg. Data courtesy of D. Stetka.[27]

medium containing BrdU. One is a method developed by Stetka[27] in which peripheral lymphocytes are collected from the ear vein of a rabbit treated with the test agent. This method has been successful with several known mutagen/carcinogens such as EMS, MCA, benzo[a]pyrene, mitomycin C, and cyclophosphamide (Figure 8.9). This method has the distinct advantage of permitting sequential samples from the same animal. Thus, each animal can serve as its own control, assuring a very sensitive assay.

In vivo SCE evaluation is not considered a routine assay, and very little data have been generated using the technique. However, it appears to be a useful probe for *in vivo* analysis because of its sensitivity and scoring precision.

In Vitro Cell Transformation

Cellular transformation probably represents the closest simulation of events which occur during the oncogenic process *in vivo*. The mutagenesis/carcinogenesis relationship assumes that genomic alteration is a critical step in the transformation process, but other studies suggest that additional molecular changes must also occur for a cell to acquire the transformed properties. The *in vitro* transformation process probably includes most of the steps. Nonmutagenic steps include promotion and fixation phenomena that stimulate the expression of the DNA lesion. Table 8.6

TABLE 8.6
Hypothetical Steps Involved in Oncogenic Transformation

Step	Comment
Initiating Event ↓	This is probably the initial event and is a mutation or DNA lesion affecting a particular gene or set of genes. This step is irreversible but requires subsequent steps for expression.
Fixation ↓	The genetic lesion requires specific metabolic processes to occur and/or cell division to fix the lesion into the genome of the cell.
Promotion/Enhancement ↓	It is assumed that specific physiological or biochemical phenomena are required to facilitate the expression of the initiating event. Other chemicals, such as phorbol esters, hydrocarbons, phenobarbital, smoke condensate, etc., have been implicated in the promotion and enhancement of cell transformation. To a certain extent these processes appear reversible.
Expression of Cell Transformation ↓	The cell acquires a new set of genetic biochemical and immunological properties. The properties give rise to the tumor properties of the particular neoplasm. All progeny cells of the tumor acquire these properties fully expressed and do not go through the transformation process.
Tumor	When a sizable population of transformed cells is formed by rapid proliferation, the neoplastic disease is identified clinically as a tumor mass.

illustrates the hypothetical processes which occur during the transformation process.

The number of genomic lesions necessary for oncogenic transformation is not known. If multiple hits are necessary to transform a cell, each lesion is a random event and a function of the dose received at each insult. This may explain the latency period associated with cancer. If the insults are separated over time, so too are the DNA lesions. Following initiation, the lesion must be fixed before a cell undergoes transformation. DNA lesions may be induced, for example, in cells which have a very low rate of DNA replication. This means that fixation may not occur until the cells are stimulated to undergo cell division by some external factors such as hormonal (natural or synthetic steroids) stimulation, differentiating stimulus, or regeneration of the tissue mass following a toxic insult. As long as the lesion is not fixed, the subsequent steps leading toward oncogenic transformation will not occur and the "potentially transformed cell" remains unexpressed for long periods of time.

While the steps shown in Figure 8.10 are consistent with much of what is known about the etiology of neoplasia, molecule mechanisms or formal proof of their involvement is presently lacking (Table 8.7). It is still generally assumed, however, that transformation assays, especially when set up to use a metabolic activation system and modified to look for promoting effects, are highly reliable models for animal carcinogenesis.

Ideally, human cells should be used as targets for transformation. Recent studies indicate that human cells can be transformed by chemicals, but the methodology needed for these assays is hardly ready to be applied in routine screening.[22] There are presently four or five systems that can operate in that capacity; they range from primary or passage cells to well-characterized cell lines. The most widely used assays are summarized in Table 8.8.

The published data base on transformation assays is small but expanding. The majority of compounds evaluations have relied upon the intrinsic metabolic capability of the target cells, since most untransformed cells are highly sensitive to the standard S9 microsome systems. The exception to this sensitivity is the BHK-21 cell, which is compatible with S9 mix.[25]

The addition of a metabolizing system to the transformation assays significantly enhances their utility in chemical screening. Recently, coupling target cells with primary hepatocyte metabolizing cells has been successfully accomplished in the SHE, BALB/c 3T3, and 10T½ assays. This development has substantially expanded the utility of transformation assays by permitting detection of compounds with strict requirements for activation.[24]

The mechanism of oncogenicity induced by a chemical may well be important in how such agents are controlled. Since it could be argued that agents which initiate tumors might have different induction kinetics than agents which are pure promoters, it is of extreme value to be able to recognize promoters among the classes of chemicals tested. The following is a list

TABLE 8.7
Cancer Cell Traits Explained by a Genotoxic Model

Oncogenic characteristics	Explained by model outlined in Table 8.6
Single-cell (clonal) original of tumors	The low frequency of mutagenic alterations induced in isolated cells.
Latency period	The requirement for multiple and independently occurring series of cellular events which occur postinitiation.
Tumor production by nonmutagenic agents	Chemicals which promote or stimulate fixation are necessary and can increase the background level of tumors. These agents do not initiate the original event.[a]

[a] Evidence for this process has been provided by Kopelovich, Bias, and Helson: Tumor promotor alone induces neoplastic transformation of fibroblasts from humans genetically predisposed to cancer, *Nature* **282**: 619–621, 1979.

TABLE 8.8

Summary of *in Vitro* Cell Transformation Assays

Target cell	Source	Key reference(s)	Features	Possible limitations
SHE	SHE fibroblast cells obtained from pooled litters which are trypsinized to obtain cell suspensions	12 13 24	Primary or low passage cells used as targets Spontaneous background virtually zero[a] Can be conducted as clonal or focus assay Assay can be conducted in about 2 weeks Nontransformed target cells nontumorigenic Most Type III foci demonstrate malignancy when transplanted to animals	Target cells must be preselected and tested. Availability of target cells not predictable Levels of induced transformation sometimes very low, making interpretation difficult Clonal assay evaluation requires considerable experience, and scoring is often highly subjective
BALB/c 3T3 and C₃H 10T½	BALB/c mouse fibroblast cells C₃H mouse fibroblast cells	14 21 11 23	Spontaneous background relatively low Induced numbers of foci generally high Conducted as focus assays Assays require about 4–6 weeks Nontransformed target cells nontumorigenic Most Type III foci demonstrate malignancy when transplanted to appropriate animals Cells from continuous line and therefore readily available and should be similar from lab to lab Scoring foci is easier than SHE	Target cells are from a continuous line, and therefore not equivalent to "normal" primary cells Spontaneous background susceptible to change unless carefully controlled

| Baby hamster BHK-21 cells | Baby hamster kidney fibroblast cells | 25 | Test can be conducted in 2 weeks
Transformation based on growth in soft agar and not morphology
Target cells are from a continuous line and readily available | Nontransformed cells have low level of tumorigenic potential
Spontaneous background generally very high
Growth in soft agar is quite variable, making test very difficult to score
Mechanism appears to be a single step. Process like mutation rather than transformation |

[a] Recent evidence suggests that a low spontaneous transformation frequency does exist for these cells.

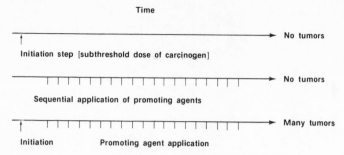

FIGURE 8.10. The effect of promotors on skin carcinogenesis in mice. One of the typical methods of carcinogenesis assessment is by skin painting. This evaluation involves chronic application of the initiating agent to mouse skin and an evaluation of the number of papillomas and carcinomas induced. Promoting agents enhance the carcinogenic effect of a single application of an initiating agent.

of agents suspected of having promotion but no initiation capability:

Pharbol esters (TPA and derivatives)   DDT
Phenobarbital   Alkanes
Butylated hydroxytoluene (BHT)   Bile salts
Anthralin

Several of these compounds have been shown to be promoters in *in vitro* transformation assays. The effects of promoters in animal skin oncogenesis are shown in Figure 8.10.

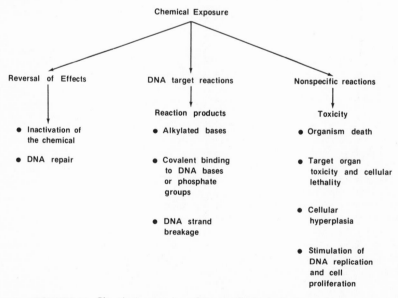

FIGURE 8.11. Chemical exposure and its possible biological consequences.

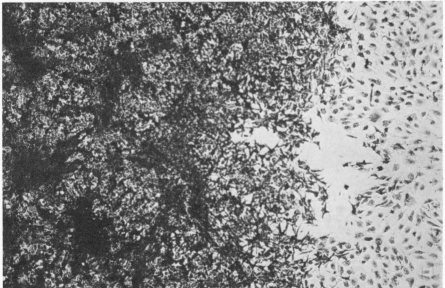

FIGURE 8.12. Cell transformation *in vitro*. The two flasks in the top portion of the figure represent treated (left) and control (right) cell populations. The large darkly stained areas are foci of transformed cells. The transformed cells are morphologically distinct from nontransformed cells as shown in the lower portion of the figure. The cells in this photograph are the Balb/c 3T3 line. Photographs courtesy of B. Myhr.

The mechanisms of tumor-promoting and tumor-enhancing agents may be quite diverse. Studies suggest that promoting agents such as TPA enhance long-lived DNA damage; reduce repair; and increase cellular proliferation, chromosomal rearrangements, and SCE frequencies. Enhancing agents are known to affect cellular transport, metabolic activation, binding of macromolecules, and cellular necrosis (Figure 8.11).

The typical BALB/c 3T3 transformation assay consists of an initial exposure of 24–72 hr followed by removal of the medium containing the test sample and biweekly media changes. At the end of 4–6 weeks, the cells have become confluent, and if the initial exposure was transforming, foci (Figure 8.12) of morphologically distinct cells are visible. A promoting study consists of repeated exposures of the suspect promoter subsequent to a subtransforming of a positive carcinogen.

The transformation assay promises to be a powerful component in the battery of short-term carcinogenicity assays.

REFERENCES

1. Allen, J. W., Shuler, C. F., Mendes, R. W. and Latt, S. A.: A simplified technique for *in vivo* analysis of sister chromatid exchanges using 5-bromodeoxyuridine tablets. *Cytogenet. Cell Genet.* **18**:231–237, 1977.
2. Ames, A. N., McCann, J., and Yamasaki, E.: Methods for detecting carcinogens and mutagens with the *Salmonella*/mammalian-microsome mutagenicity test. *Mutat. Res.* **31**:347–364, 1975.
3. Bignami, M., Morpurgo, G., Agliani, R., Carere, A., Conti, G., and Di Giuseppe, G.: Non-disjunction and crossing-over induced by pharmaceutic drugs in *Aspergillus nidulans*. *Mutat. Res.* **26**:159–170, 1974.
4. Boyd, J. B. and Setlow, R. B.: Characterization of post-replication repair in mutagen-sensitive strains of *Drosophila melanogaster*. *Genetics* **84**:507–526, 1976.
5. Brown, M. M., Wassom, J. S., Malling, H. V., Shelby, M. D. and Von Halle, E. S.: Literature survey of bacterial, fungal, and *Drosophila* assay systems used in the evaluation of selected chemical compounds for mutagenic activity. *J. Natl. Cancer Inst.* **62**(4):841–871, 1979.
6. Brusick, D.: Bacterial mutagenesis and its role in the identification of potential animal carcinogens. In *Carcinogens: Identification and Mechanisms of Action* (A. C. Griffin and C. R. Shaw, eds.), Raven Press, New York, pp. 93–105, 1979.
7. Brusick, D. J.: Induction of cycloheximide-resistant mutants in *Saccharomyces cerevisiae* with *N*-methyl-*N'*-nitro-*N*-nitrosoguanidine and ICR-170. *J. Bacteriol.* **109**(3):1134–1138, 1972.
8. Brusick, D. J.: Observations and recommendations regarding routine use of bacterial mutagenesis assays as indicators of potential chemical carcinogens. In *Strategies for Short-Term Testing for Mutagens/Carcinogens* (Byron E. Butterworth, ed.), CRC Press, West Palm Beach, Fla., pp. 3–12, 1979.
9. Callen, D. F., and Philpot, R. M.: Cytochrome P-450 and the activation of promutagens in *Saccharomyces cerevisiae*. *Mutation Res.* **45**:309–324, 1977.
10. Carrano, A. V., Thompson, L. H., Lindl, P. A., and Minkler, J. L.: Sister chromatid exchange as an indicator of mutagenesis. *Nature* (*London*) **271**:551–553, 1978.

11. Chen, T. T., and Heidelberger, C.: Quantitative studies on the malignant transformation *in vitro* of cells derived from adult C3H mouse ventral prostate. *Int. J. Cancer* **4**:166–167, 1969.

12. DiPaolo, J. A., Donovan, P., and Nelson, R. L.: Quantitative studies of *in vitro* transformation by chemical carcinogens. *J. Natl. Cancer Inst.* **42**:867–874, 1969.

13. DiPaolo, J. A., Nelson, R. L., and Donovan, P. J.: Morphological, oncogenic, and karyological characteristics of Syrian hamster embryo cells transformed *in vitro* by carcinogenic polycyclic hydrocarbons. *Cancer Res.* **31**:1118–1127, 1971.

14. DiPaolo, J. A., Takano, K., and Popescu, N. C.: Quantitation of chemically induced neoplastic transformation of BALB/3T3 cloned cell lines. *Cancer Res.* **32**:2686–2695, 1972.

15. Ehling, U. H.: Specific-locus mutations in mice. In *Chemical Mutagens: Principles and Methods for Their Detection*, Vol. 5 (A. Hollaender and F. J. de Serres, eds.), Plenum Press, New York, pp. 233–256, 1978.

16. Ehling, U. H., Machemer, L., Buselmaier, W., Dýcka, J., Frohberg, H., Kratochvilova, J., Lang, R., Lorke, D., Müller, D., Peh, J., Rohrborn, G., Roll, R., Schulze-Schencking, M., and Wiemann, H.: Standard protocol for the dominant lethal test on male mice. Set up by the work group, Dominant Lethal Mutations of the *Ad Hoc* Committee on Chemogenetics. *Arch. Toxicol.* **39**:173–185, 1978.

17. Fahrig, R.: The mammalian spot test: a sensitive *in vivo* method for the detection of genetic alterations in somatic cells of mice. In *Chemical Mutagens: Principles and Methods for Their Detection*, Vol. 5 (A. Hollaender and F. J. de Serres, eds.), Plenum Press, New York, pp. 151–176, 1978.

18. Generoso, W. M., Cain, K. T., Huff, S. W., and Gosslee, D. G.: Heritable translocation test in mice. In *Chemical Mutagens: Principles and Methods for Their Detection*, Vol. 5 (A. Hollaender and F. J. de Serres, eds.), Plenum Press, New York, 1978.

19. Hollstein, M., McCann, J., Angelosanto, F., and Nichols, W.: Short-term tests for carcinogens and mutagens. *Mutat. Res.* **65**:133–226, 1979.

20. Kada, T., Hirano, K., and Shirasu, Y.: Screening of environmental chemical mutagens by the Rec-assay system with *Bacillus subtilis*. In *Chemical Mutagens: Principles and Methods for Their Detection*, Vol. 6 (F. J. de Serres and A. Hollaender, eds.), pp. 149–173, Plenum Press, New York, 1980.

21. Kakunaga, T.: Quantitative system for assay of malignant transformation by chemical carcinogens using a clone derived from BALB/3T3. *Int. J. Cancer* **12**:463–473, 1973.

22. Kakunaga, T.: The transformation of human diploid cells by chemical carcinogens. In *Origins of Human Cancer*, Vol. 4 (H. H. Hiatt, J. D. Watson, and J. A. Winsten, eds.), Cold Spring Harbor Conferences on Cell Proliferation, Cold Spring Harbor Laboratory, Cold Spring Harbor, N.Y., pp. 1537–1548, 1977.

23. Marquardt, H., Juroki, T., Huberman, E., Selkirk, J. K., Heidelberger, C., Grover, P. L., and Sims, P.: Malignant transformation of cells derived from mouse prostate by epoxides and other derivatives of polycyclic hydrocarbons. *Cancer Res.* **32**:716–720, 1972.

24. Pienta, R. J.: A hamster embryo cell model system for identifying carcinogens. In *Carcinogens: Identification and Mechanisms of Action*, A. C. Griffin and C. R. Shaw, eds.), Raven Press, New York, pp. 121–141, 1979.

25. Purchase, I. F. H., Longstaff, E., Ashby, J., Styles, J. A., Anderson, D., Lefevre, P. A., and Westwood, F. R.: Evaluation of six short term tests for detecting organic chemical carcinogens and recommendations for their use. *Nature (London)* **264**:624–627, 1976.

26. Rosenkranz, H. S., and Leifer, Z.: Determining the DNA-modifying activity of chemicals using DNA-polymerase-deficient *Escherichia coli*. In *Chemical Mutagens: Principles and Methods for Their Detection*, Vol. 6 (F. J. de Serres and A. Hollaender, eds.), pp. 109–147, Plenum Press, New York, 1980.

27. Stetka, D. G.: Sister chromatid exchange in animals exposed to mutagenic carcinogens—

an assay for genetic damage in humans. Proceedings of the 9th Annual Conference on Environmental Toxicology. Dayton, Oh., Aerospace Medical Research Laboratory, AMRL-TR-79-68, 1979, pp. 215–222.

28. Swenberg, J. A., and Petzold, G. L.: The usefulness of DNA damage and repair assays for predicting carcinogenic potential of chemicals. In *Strategies for Short-Term Testing for Mutagens/Carcinogens* (Byron E. Butterworth, ed.), CRC Press, West Palm Beach, Fla, pp. 77–86, 1979.

29. Taylor, J. H.: Sister chromatid exchange in tritium-labeled chromosomes. *Genetics* **43**:515–529, 1958.

30. Vogel, E.: Identification of carcinogens by mutagen testing in *Drosophila:* The relative reliability for the kinds of genetic damage measured. In *Origins of Human Cancer*, Vol. C, Cold Spring Harbor Conferences on Cell Proliferation, Cold Spring Harbor Laboratory, Cold Spring Harbor, N.Y., pp. 1483–1497, 1977.

31. Vogel, E., and Sobels, F. H.: The function of *Drosophila* in genetic toxicology testing. In *Chemical Mutagens: Principles and Methods for Their Detection*, Vol. 4 (A Hollaender, ed.), Plenum Press, New York, pp. 93–142, 1976.

32. Wiemann, H., and Lang, R.: Strategies for detecting heritable translocations in male mice by fertility testing. *Mutat. Res.* **53**:317–326, 1978.

33. Williams, G. M.: The detection of chemical mutagens/carcinogens by DNA repair and mutagenesis in liver cultures. In *Chemical Mutagens: Principles and Methods for Their Detection*, Vol. 6 (F. J. de Serres and A. Hollaender, eds.), pp. 61–79, Plenum Press, New York, 1980.

34. Würgler, F. E., Sobels, F. H., and Vogel, E.: *Drosophila* as assay system for detecting genetic changes. In *Handbook of Mutagenicity Test Procedures*, (B. J. Kilbey, M. Legator, W. Nichols, and C. Ramel, eds.), Elsevier/North-Holland, Amsterdam, pp. 335–373, 1977.

35. Zimmermann, F. K.: Procedures used in the induction of mitotic recombination and mutation in the yeast *Saccharomyces cerevisiae*. *Mutat. Res.* **31**:71–86, 1975.

Sample Study Designs

SELECTED STUDY DESIGNS

The following compilation of study designs is intended as a resource for both reviewers and practitioners of genetic toxicology. The study designs include the general requirements under Good Laboratory Practices regulations, methodology, and evaluation criteria. Other study designs using other target organisms could be included in addition to or as replacements for those presented. The group chosen represents a broad range of studies and provides the basis for the development of a sound genetic toxicology program for either screening or risk estimation.

These study designs are all in current use and therefore have been through the processes of validation. The sizes of the experiments described in the study designs were selected on the basis of routine screening. Increases or decreases in the magnitude of each study can be made to suit special situations. Each study design is provided with relevant references to more detailed discussions of the technique.

PROTOCOL 1

AMES SALMONELLA/MICROSOME PLATE ASSAY

The objective of this study is to evaluate a test article for mutagenic activity in a bacterial assay with and without a mammalian S9 activation system.

Materials

The following microbial strains of *S. typhimurium* should be routinely used in the plate test:

| Strain designation | Gene affected | Additional mutations | | | Mutation type detected |
		Repair	LPS	R Factor	
TA-1535	*hisG*	Δ*uvrB*	*rfa*	—	Base-pair substitution
TA-1537	*hisC*	Δ*uvrB*	*rfa*	—	Frameshift
TA-1538	*hisD*	Δ*uvrB*	*rfa*	—	Frameshift
TA-98	*hisD*	Δ*uvrB*	*rfa*	pKM101	Frameshift
TA-100	*hisG*	Δ*uvrB*	*rfa*	pKM101	Base-pair substitution

The *Salmonella* cultures can be obtained from Dr. B. N. Ames University of California at Berkeley.

Each strain should be cultured in nutrient broth for approximately 16 hr at 37°C (titer of 10^8–10^9 cells/ml) and used for the mutagenicity test. Examine the master culture of each strain for appropriate genetic markers (*his, uvrB, rfa,* pkM101). The data of culture preparation should be documented and the daily cultures should be examined periodically for appropriate genetic markers as described by Ames, *et al.* (1975).

An S9 homogenate should be used as the activation system (see Appendix A). The 9000*g* supernatant is prepared from Sprague–Dawley adult male rat liver induced by Aroclor 1254 (described by Ames *et al.*, 1975). The S9 mix should have the following components:

Components	Concentration/ml S9 mix
NADP (sodium salt)	4 μmol
D-Glucose-6-phosphate	5 μmol
MgCl₂	8 μmol
KCl	33 μmol
Sodium phosphate buffer, pH 7.4	100 μmol
Organ homogenate from rat liver (S9 fraction)	100 μl

The bacterial strains should be cultured in Oxoid Media #2 (Nutrient Broth), with selective medium Vogel Bonner Medium E with 2% glucose. According to the methods of Ames *et al.* (1975), the overlay agar should consist of 0.6% purified agar with 0.05 mM histidine, 0.05 mM biotin, and 0.1 M NaCl.

Test Procedures

Prepare and label an appropriate number of tubes of molten overlay agar as well as Vogel Bonner plates.

To set up the stock solutions, the test article should be weighed or measured and diluted into a solvent. The test article and other components should be added to the overlay and the contents poured on the surfaces of the appropriate Vogel Bonner plates (see Figure 9.1).

The plates (once the overlay has solidified) should be incubated at 37°C for 36–72 hr and then scored for numbers of revertants per plate.

All tests should be evaluated using a minimum of five concentrations. In the standard plate test, at least seven dose levels of the test article, dissolved in a suitable solvent, should be added to the test system. This step allows for high dose toxicity without loss of the entire test. One to three plates per dose per strain should be used in the standard assay. The standard test doses should be 0.005, 0.01, 0.1, 1.0, 5.0, and 10.0 μl per plate for liquids; and 0.5, 1.0, 10.0, 100.0, 500.0, and 1000.0 μg per plate for solids. Additional lower doses should be employed in the tests if toxicity is observed at the three highest doses. When no toxicity is observed, additional

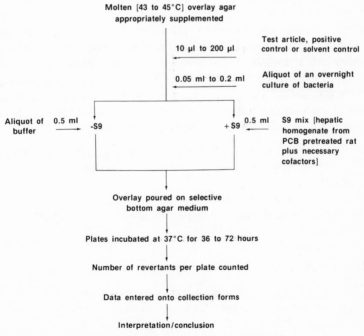

FIGURE 9.1. Reverse Mutation Assay in Bacteria.

concentrations may be employed up to 50 μl or 5000 μg per plate. Alternatively to a preset concentration range one can elect to perform range-finding tests. A preliminary toxicity range-finding test with one strain such as TA-100 may be helpful in establishing a correct range of concentrations. The LC50 is appropriate as a high dose based on a toxicity test.

The articles that should be used as positive control agents are:

Assay	Chemical	Solvent	Concentration per plate (μg)	Responsive *Salmonella* strains
Nonactivation	Sodium azide	Water	1	TA-1535, TA-100
	2-nitrofluorene (NF)	Dimethyl-sulfoxide	10	TA-1538, TA-98
	9-aminoacridine (9AA)	Ethanol	50	TA-1537
Activation	2-anthramine (ANTH)	Dimethyl-sulfoxide	2.5	For all strains

Evaluation Criteria

Since statistical methods are not currently used, evaluation can be based on the criteria included in this protocol.

Plate test data will consist of direct revertant colony counts obtained from a set of selective agar plates seeded with populations of mutant cells suspended in a semisolid overlay. Because the test article and the cells are incubated in the overlay for 36–72 hr and a few cell divisions occur during the incubation period, the test is semiquantitative in nature. Although these features of the assay reduce the quantitation of results, they provide certain advantages not contained in a quantitative suspension test:

1. The small number of cell divisions permits potential mutagens to act on replicating DNA, which is often more sensitive than nonreplicating DNA.
2. The combined incubation of the test article and the cells in the overlay permits constant exposure of the indicator cells for 36–72 hr.

Surviving Populations

Plate test procedures do not permit exact quantitation of the number of cells surviving chemical treatment. At low concentrations of the test article, the surviving populations on the treatment plates are essentially the same as that on the negative control plate. At high concentrations, the surviving populations are usually reduced by some fraction. This protocol normally employs several doses ranging over two or three log concentrations; the highest of these doses is selected to show slight toxicity as determined by subjective criteria such as a slight reduction in the number of background revertants per plate.

Dose–Response Phenomena

The demonstration of dose-related increases in mutant counts is an important criterion in establishing mutagenicity. A factor that might modify dose-response results for a mutagen would be the selection of doses that are too low (usually mutagenicity and toxicity are related). If the highest dose is far lower than a toxic concentration, no increases may be observed over the dose range selected. Conversely, if the lowest dose employed is highly cytotoxic, the test article may kill any mutants that are induced, and thus will not appear to be mutagenic.

Control Tests

Positive and negative control assays should be conducted with each experiment and should consist of direct-acting mutagens for nonactivation assays and mutagens that require metabolic biotransformation in activation assays. Negative controls should consist of the test article solvent in the overlay agar together with the other essential components. The negative control plate for each strain gives a reference point to which the test data can be compared. The positive control assay is conducted to demonstrate that the test systems are functional with known mutagens.

Evaluation Criteria for Ames Assay

Because the procedures to be used to evaluate the mutagenicity of the test article are semiquantitative, the criteria used to determine positive effects are inherently subjective and based primarily on a historical data base. Most data sets should be evaluated using the following criteria.

1. *Strains TA-1535, TA-1537, and TA-1538.* If the solvent control value is within the normal range, a test article that produces a positive dose response over *three* concentrations, with the highest increase equal to three times the solvent control value, should be considered mutagenic.

2. *Strains TA-98 and TA-100.* If the solvent control value is within the normal range, a test article that produces a positive dose response over *three* concentrations, with the highest increase equal to twice the solvent control value for TA-98 and TA-100, should be considered mutagenic.

3. *Pattern.* Because TA-1535 and TA-100 are derived from the same parental strain (G-46) and because TA-1538 and TA-98 are derived from the same parental strain (D3052), to some extent there is a built-in redundancy in the microbial assay. In general, the two strains of a set respond to

the same mutagen, and such a pattern is sought. Generally, if a strain responds to a mutagen in nonactivation tests, it should do so in activation tests.

4. *Reproducibility*. If a test article produces a response in a single test that cannot be reproduced in additional runs, the initial positive test data lose significance.

The preceding criteria are not absolute, and other extenuating factors may enter into a final evaluation decision. However, these criteria can be applied to the majority of situations and are presented to aid those individuals not familiar with this procedure. A normal, spontaneous background range for a laboratory should be documented and compared to published ranges.

Relation between Mutagenicity and Carcinogenicity

It must be emphasized that the Ames *Salmonella*/microsome plate assay is not a definitive test for chemical carcinogens. It is recognized, however, that correlative and functional relations have been demonstrated between these two end points. The results of comparative tests on 300 chemicals by McCann *et al.* (1975) showed an extremely good correlation between results of microbial mutagenesis tests and *in vivo* rodent carcinogenesis assays. Modifications of the procedure involving preincubation conditions or sources of S9 mix are nessary for evaluation of specific chemicals or classes of chemicals.

References

Ames, B. N., McCann, J., and Yamasaki, E.: Methods for detecting carcinogens and mutagens with the *Salmonella*/mammalian microsome mutagenicity test. *Mutat. Res.* **31**:347–364, 1975.
McCann J., Choi, E., Yamasaki, E., and Ames, B. N.: Detection of carcinogens as mutagens in the *Salmonella*/microsome test: Assay of 300 chemicals. *Proc. Natl. Acad. Sci. U.S.A.* **72**:5135–5139, 1975.

PROTOCOL 2

MITOTIC RECOMBINATION IN *S. CEREVISIAE* STRAIN D₃

The objective of this study is to evaluate a test article for its genetic activity in *S. cerevisiae* yeast strain D_3 with and without the addition of metabolic activation preparations (S9).

It has been found that some environmental chemicals are preferentially active in eukaryotic cells. Assays employing these cells, coupled with microbial bioassays, can be effective prescreens to assess whether a test article should be examined in more definitive tests. The yeast strain used in this study is a diploid eukaryotic organism in which the induction of mitotic recombination by a test article can be measured.

Normally, recombination is measured in meiotic cells, but under defined conditions it can be detected in mitotically dividing cells. The induction of mitotic recombination is measured by the production of homozygous alleles from heterozygous alleles (see Chapter 8).

Dosing Procedure

All tests should evaluated using at a minimum of four concentrations. At least six dose levels of the test article dissolved in a suitable solvent should be added to the test system. The standard test doses should be 0.005, 0.01, 0.1, 1.0, 5.0, and 10.0 μl per plate for liquids; and 0.5, 1.0, 10.0, 100.0, 500.0, and 1000.0 μg per plate for solids. Additional lower doses should be employed in the tests if toxicity is observed at the three highest doses. A preliminary toxicity test can be performed on a test article; based on the results, four dose levels can be selected for the assay. An LC50 from the toxicity test is appropriate as the high dose.

An S9 homogenate should be used as the activation system. The 9000g supernatant is prepared from Sprague–Dawley adult male rat liver induced by Aroclor 1254 (described by Ames *et al.*, 1975) (see Appendix A).

Method

Cell cultures for this study should be prepared from a single isolated colony of the indicator organism grown on yeast complete agar. The isolate should be suspended in saline, and 0.2 ml of the suspension spread onto complete medium and grown at 30° for 2 days. (See Brusick and Mayer, 1973, for media preparation.) The cell lawn should then be collected into a suspension using saline. A cell titer of 1 ml should be prepared. Approximately 10^7 cells from the suspension stock of the indicator strain should be added to a screw-capped test tube. For the nonactivation test, 0.5 ml of phosphate buffer, pH 7.4, should be added to each tube. In activation test, an aliquot of reaction mixture (0.5 ml containing 9000g liver homogenate) should be added to each tube in place of the buffer. Six concentration levels of the test article should be added to the appropriate tubes. Incubate all tubes at 30°C in a shaker water bath for 3 hr. Remove samples dilute in 0.15 M saline, and plate onto complete medium. All plates should be incubated at 30°C for 4 days. The plates should be screened for pigmented colonies and sectors using a dissecting microscope with variable magnification. Positive and solvent controls using both directly active positive chemicals and those requiring metabolic activation should be run with each assay. EMS and DMN are suitable as positive criteria. Scored after 3–5 days of incubation, plates with suitable densities of colonies (500–1000) are examined under a dissecting microscope for the presence of red sectors in the normally white colonies. Occasionally the entire colony will be red. The frequency of red sectors or colonies is calculated for each control and test group. A total of 15,000–20,000 colonies should be scored per group.

Evaluation Criteria

The yeast strain D_3 is a diploid strain of *S. cerevisiae* with the genotype *CYH4*/+, *ade2*/+. *his8*/+, used for studying genetic events such as reciprocal and nonreciprocal mitotic recombination.

Though the reciprocal recombination, which involves the exchange of genetic material between two nonsister chromatids during mitosis, is a nonmutational genetic event, it results in the homozygosity of the recessive genes. This homozygosity could bring about deleterious phenotypes, which are an indication of chromosomal and, hence, DNA alteration. The phenotypic expression (or the events) of reciprocal recombination in yeast strain D_3 are the red sectored cells representing the homozygosity of the two recessive *ade2* alleles. A concomitant homozygosity of histidine (*his8*) can be used to confirm the event as a crossover between the centromere and the *ade2* locus. The presence of all red colonies can also be used as evidence for mitotic recombination or gene conversion.

These recombinogenic events that occur in mitosis are rare; nonetheless, when they do occur, they can lead to deleterious effects. The results could be considered positive if the total events in a test are equal to or greater than two times the spontaneous events. An accompanying dose-related effect is also necessary to give confidence to the increase.

Positive and negative control assays should be conducted with each experiment, and consist of direct-acting mutagens for nonactivation assays and mutagens that require metabolic biotransformation for activation assays. Negative controls should consist of the test article solvent with the other essential components. The negative control data give a reference point to which the test data can be compared. The positive control assay is conducted to demonstrate that the test systems are functional with known mutagens.

References

Ames, B. N., McCann, J., and Yamasaki, E.: Methods for detecting carcinogens and mutagens with the *Salmonella*/mammalian microsome mutagenicity test. *Mutat. Res.* **31**:347–364, 1975.

Brusick, D. J., and Mayer, V. W.: New developments in mutagenicity screening techniques with yeast. *Environ. Health Perspect.* **6**:83–96, 1973.

Zimmermann, F. K.: A yeast strain for visual screening for the two reciprocal products of mitotic crossing over. *Mutat. Res.* **21**:263–269, 1973.

Zimmermann, F. K., Kern, R., and Rasenberger, H.: A yeast strain for simultaneous detection of induced mitotic crossing over, mitotic gene conversion and reverse mutation. *Mutat. Res.*, **28**:381–388, 1975.

PROTOCOL 3

MITOTIC GENE CONVERSION IN *S. CEREVISIAE* STRAIN D_4

The objective of this study is to evaluate a test article for its genetic activity in the yeast strain D_4 with and without the addition of mammalian metabolic activation system using preincubation treatment conditions. D_4 is a diploid strain of *S. cerevisiae*, heteroallelic at the adenine 2 and tryptophan 5 locus. These alleles are stable and show low frequencies of revertibility. Both heteroallelic loci result in nutritional deficiencies (noncomplementing), prohibiting the cells from growing on either minimal or single supplemented media. Mitotic gene conversion is a nonreciprocal event, probably involving alteration of a small number of nucleotide pairs within a single gene, that can generate a wild-type allele at either of the heteroallelic sites. This results in the expression of a functional gene and the loss of the nutritional requirement.

Dosing Procedure

All tests should be evaluated using a minimum of four concentrations. At least six dose levels of the test article dissolved in a suitable solvent should be added to the test system. Standard test doses should be 0.005, 0.01, 0.1, 1.0, 5.0, and 10.0 μl per plate for liquids; and 0.5, 1.0, 10.0, 100.00, 500.0, and 1000.0 μg per plate for solids. Additional lower doses should be employed in the tests if toxicity is observed at the three highest doses. A preliminary toxicity test can be performed on the test article; based on the results, four dose levels can be selected for the assay using the LC50 as the high dose.

An S9 homogenate should be used as the activation system. The 9000g supernatant is prepared from Sprague–Dawley adult male rat liver induced by Aroclor 1254 (described by Ames *et al.*, 1975) (see Appendix A).

Method

Add approximately 10^7 cells (0.1 ml) of the indicator strain from a culture grown overnight to test tubes. Six dose levels of the test article and 0.5 ml of S9 mix (for activation assay) or 0.5 ml phosphate buffer, pH 7.4 (for nonactivation assay), should be added to the contents of the appropriate tubes. Mix the contents of the tubes gently for 3 hr at 37°C. Then, 2.0 ml of molten overlay agar (supplemented with 0.05 mM tryptophan) should be added, the contents of the tubes mixed, and the mixture poured onto minimal agar plates (see Brusick and Andrews, 1975, for medium preparation). Incubate the plates at 30°C for 3–5 days and then score. Plates are scored for the number of colonies growing per petri dish. One to three plates per dose can be used. Positive and solvent controls using both directly active positive chemicals and those that require metabolic activation should be run with each assay. EMS and 2-acetylaminofluorene (2-AAF) can be used as positive controls.

General Evaluation Criteria

Plate test data consist of direct convertant colony counts obtained from a set of selective agar plates seeded with populations of cells suspended in a semisolid overlay. Because the test article and cells are incubated in the overlay for 3–5 days, and a few cell divisions occur during the incubation period, the test is semiquantitative in nature. Althought these features of the assay reduce the quantitation of results, they provide certain advantages not contained in a quantitative suspension test:

1. The small number of cell divisions permits potential mutagens to act on replicating DNA, which is often more sensitive than nonreplicating DNA.
2. The combined incubation of the test article and the cells in the overlay permits constant exposure of the indicator cells to test article for 3–5 days.

Surviving Populations

Plate test procedures do not permit exact quantitation of the number of cells surviving test article treatment. At low concentrations of the test article, the surviving population on the treatment plates is essentially the same as that on the negative control plate. At high concentrations, the surviving population is usually reduced by some fraction. This protocol normally employs several doses ranging over two or three log concentrations; the highest of these doses is selected to show slight toxicity as determined by subjective criteria.

Dose–Response Phenomena

The demonstration of dose-related increases in convertant counts is an important criterion in establishing mutagenicity. A factor that might modify dose-response results for a mutagen would be the selection of doses that are too low (usually, mutagenicity and toxicity are related). If the highest dose is far lower than a toxic concentration, no increases may be observed over the dose range selected. Conversely, if the lowest dose employed is highly cytotoxic, the test article may kill any convertants that are induced, and thus will not appear to be mutagenic.

Control Tests

Positive and negative control assays should be conducted with each experiment and should consist of direct-acting mutagens for nonactivation assays and mutagens that require metabolic biotransformation in activation assays. Negative controls should consist of the test article solvent in the overlay agar together with the other essential components. The negative control plate for each strain gives a reference point to which the test data can be compared. The positive control assay is conducted to demonstrate that the test systems are functional with known mutagens.

Evaluation Criteria for the Preincubation Plate Incorporation Assays Using Yeast Strain D_4

Because the procedures used to evaluate the mutagenicity of the test article are semiquantitative, the criteria used to determine positive effects are inherently subjective and based primarily on a historical data base. Basically, it the solvent control value is within the normal range, a test article that produces a positive dose response over *three* concentrations, with the highest increase equal to twice the solvent control value, should be considered as an agent inducing mitotic gene conversions.

Reproducibility

If a test article produces a response in a single test that cannot be reproduced in one or more additional runs, the initial positive test data lose significance.

The preceding criteria are not absolute, and other extenuating factors may enter into a final evaluation decision. However, these criteria are applied to the majority of situations and are presented to aid those individuals not familiar with this procedure. As the data base is increased, the criteria for evaluation can be more firmly established.

All evaluations and interpretation of the data should be based only on the demonstration, or lack, of mutagenic activity.

References

Ames, B. N., McCann, J., and Yamasaki, E.: Methods for detecting carcinogens and mutagens with the *Salmonella*/mammalian microsome mutagenicity test. *Mutat. Res.* **31**:347–364, 1975.

Brusick, D., and Andrews, H.: Comparison of the genetic activity of dimethylnitrosamine ethylmethane sulfonate, 2-acetylaminofluorene and ICR-170 in *Saccharomyces cerevisiae* strains D_3, D_4 and D_5 using *in vitro* assays with and without metabolic activation. *Mutat. Res.* **26**:491–500, 1974.

PROTOCOL 4

MITOTIC RECOMBINATION IN *S. CEREVISIAE* STRAIN D_5

The objective of this study is to evaluate a test article for its genetic activity in *S. cerevisiae* yeast strain D_5 with and without the addition of metabolic activation preparations (S9). Strain D_5 is similar to strain D_3 with the exception that both products of reciprocal recombination in strain D_5 can be visualized in the affected colony.

Dosing Procedure

All tests should be run at a minimum of four concentrations. At least six dose levels of the test article dissolved in a suitable solvent should be added to the test system. The standard test doses should be 0.005, 0.01, 0.1, 1.0, 5.0, and 10.0 μl per plate for liquids; and 0.5, 1.0, 10.0, 100.0, 500.0, and 1000.0 μg per plate for solids. Additional doses may be employed in the tests if toxicity is observed at the three highest doses. A preliminary toxicity test can be performed on a test article; based on the results, four dose levels can be selected for the assay.

An S9 homogenate should be used as the activation system. The 9000g supernatant is prepared from Sprague–Dawley adult male rat liver induced by Aroclor 1254 (decribed by Ames *et al.*, 1975) (see Appendix A).

Method

Cell cultures for this study should be prepared from a single isolated colony of the indicator organism. The isolate should be suspended in saline, and 0.2 ml of the suspension spread onto complete medium (see Brusick and Mayer, 1973, for medium preparation). The lawns should then be collected into a suspension using saline. Approximately 10^7 cells from the suspension stock of the indicator strain should be added to a screw-capped test tube. For the nonactivation test, 0.5 ml of phosphate buffer, pH 7.4, should be added to each tube. In the activation test, an aliquot of reaction mixture (0.5 ml containing 9,000g liver homogenate) should be added to each tube in place of the buffer. At least six concentration levels of the test article should be added to the appropriate tubes. All tubes should be incubated at 30°C with gentle agitation in a shaker water bath for 3 hr. Samples should then be removed, diluted in 0.15 M saline, and plated onto complete medium. All plates should be incubated at 30°C for 4 days. The plates should be screened for pigmented colonies and sectors using a dissecting microscope with variable magnification. Positive and solvent controls using both directly active positive chemicals and those requiring metabolic activation should be run with each assay. EMS and DMN are suitable positive control agents for this assay.

Evaluation Criteria

The yeast strain D_5 is a diploid strain of S. cerevisiae with the genotype ade2-40/+, +/ade2-119 used for studying genetic events such as reciprocal and nonreciprocal mitotic recombination. The complementing adenine, ade2 markers (ade2-40 and ade2-119), are used for monitoring the recombinational events in the strain D_5. Approximately 15–20,000 colonies should be scored per treatment or control level.

Though the reciprocal recombination, which involves the exchange of genetic material between two nonsister chromatids during mitosis, is a nonmutational genetic event, it results in the homozygosity of the recessive genes. This homozygosity could bring about deleterious phenotypes that are an indication of chromosomal and, hence, DNA alteration. The phenotypic expression (or the events) of reciprocal recombination in yeast strain D_5 are the red-pink cells representing the homozygosity of the two recessive ade2 alleles—ade2-40 and ade2-119. The nonreciprocal recombination, which is also known as gene conversion, is again a nonmutational genetic event that can occur in dividing or resting cells. At the two-strand stage of nondividing cells of this diploid strain, the nonreciprocal recombination forms red or pink colonies; at the four-strand replicative stage during cell division, gene conversion brings about red-white or pink-white sectored colonies. A full description of the development of sector colonies is given by Zimmerman (1973).

These recombinogenic events that occur in mitosis are rare; nonetheless, when they do occur, they can lead to deleterious effects. The results could be considered as positive if the total events in a test are equal to or greater than two times the spontaneous events. An accompanying dose-related effect is also necessary to give confidence to the increase. Plate assays are required to demonstrate clear dose-related increases in mutant counts over at least three consecutive concentrations to be considered positive.

Positive and negative control assays should be conducted with each experiment, and consist of direct-acting mutagens for nonactivation assays and mutagens that require metabolic biotransformation for activation assays. Negative controls should consist of the test article solvent with the other essential components. The negative control data give a reference point to which the test data can be compared. The positive control assay is conducted to demonstrate that the test systems are functional with known mutagens.

These criteria have been developed to provide information regarding the approach taken in evaluating responses. They are not meant to be absolute and may vary if circumstances warrant.

References

Ames, B. N., McCann, J., and Yamasaki, E.: Methods for detecting carcinogens and mutagens with the *Salmonella*/mammalian microsome mutagenicity test. *Mutat. Res.* **31**:347–364, 1975.

Brusick, D. J., and Mayer, V. W.: New developments in mutagenicity screening techniques with yeast. *Environ. Health Perspect.* **6**:83–96, 1973.

Zimmermann, F. K.: A yeast strain for visual screening for the two reciprocal products of mitotic crossing over. *Mutat. Res.* **21**:263–269, 1973.

Zimmermann, F. K., Kern, R., and Rasenberger, H.: A yeast strain for simultaneous detection of induced mitotic crossing over, mitotic gene conversion and reverse mutation. *Mutat. Res.* **28**:381–388, 1975.

PROTOCOL 5

L5178Y TK$^{+/-}$ MOUSE LYMPHOMA FORWARD MUTATION ASSAY

This assay evaluates a test article for its ability to induce forward mutation in the L5178Y TK$^{+/-}$ mouse lymphoma cell line, as assessed by colony growth in the presence of BrdU or TFT. The test can also be used to confirm, in mammalian cells, presumptive mutagenicity demonstrated in microbial screens.

TK is a cellular enzyme that allows cells to salvage thymidine from the surrounding medium for use in DNA synthesis. If a thymidine analog such as BrdU is included in the growth medium, the analog will be phosphorylated via the TK pathway and incorporated into DNA, eventually resulting in cellular death. Cells which are heterozygous at the *TK* locus (*TK*$^{+/-}$) may undergo a single step forward mutation to the *TK*$^{-/-}$ genotype in which little or no TK activity remains. Such mutants are as viable as the heterozygotes in normal medium because DNA synthesis proceeds by *de novo* synthetic pathways that do not involve thymidine as an intermediate. The basis for selection of the *TK*$^{-/-}$ mutants is the lack of any ability to utilize toxic analogs of thymidine, which enable only *TK*$^{-/-}$ mutants to grow in the presence of BrdU. Cells which grow to form colonies in the presence of BrdU are therefore assumed to have mutated, either spontaneously or by the action of a test article, to the $TK^{-/-}$ genotype. These steps are illustrated diagrammatically in Figure 8.2c.

Materials

Indicator Cells

The mouse lymphoma cell line used in this assay, L5178Y $TK^{+/-}$, is derived from the Fischer L5178Y line. Stocks should be maintained in liquid nitrogen, and laboratory cultures periodically checked for the absence of mycoplasma contamination by culturing methods. To reduce the negative control frequency (spontaneous frequency) of $TK^{-/-}$ mutants to as low a level as possible, cell cultures should be exposed to conditions which select against the $TK^{-/-}$ phenotype and then return to normal growth medium for 3 or more days before use. Cells can be obtained from Dr. Donald Clive at Boroughs Wellcome Laboratories, Research Triangle Park, North Carolina.

Media

The cells should be maintained in Fischer's mouse leukemia medium supplemented with L-glutamine, sodium pyruvate, and horse serum (10% by volume). Cloning medium should consist of the preceding growth medium with the addition of agar to a final concentration of 0.35% to achieve a semisolid state. Selection medium should be cloning medium containing 100 μg/ml of BrdU or 3 μg/ml of TFT.

Control Articles

A negative control consiting of assay procedures performed on untreated cells should be utilized in all cases. If the test article is not soluble in growth medium, an organic solvent [normally dimethylsulfoxide (DMSO)] should be used; the final concentration of solvent in the growth medium should be 1% or less. Cells exposed to solvent in the medium should also be assayed as the solvent negative control article to determine any effects on survival or mutation caused by the solvent alone. For test articles assayed with activation, the untreated and solvent negative control articles should include the activation mixture.

EMS is highly mutagenic via alkylation of cellular DNA and can be used at 0.5 μl/ml as a positive control article for nonactivation studies.

DMN requires metabolic activation by microsomal enzymes to become mutagenic and can be used at 0.3 μl/ml as a positive control article for assays performed with activation.

Sample Forms

All types of test articles can be evaluated in the mouse lymphoma assay. Solid articles can be dissolved in growth medium, if possible, or in

DMSO, unless another solvent is requested. Liquids can be tested by direct addition to the test system at predetermined concentrations or following dilution in a suitable solvent. Highly volatile liquids which must be tested in the vapor phase can be added to an airtight container of fixed volume and allowed to completely volatilize in the presence of the exposed cell population. Gases can be treated by measuring known volumes of gas into an airtight container of fixed volume.

Experimental Design

Dose Selection

The solubility of the test article in growth medium and/or DMSO should be determined first. Then, a wide range of test article concentrations should be tested for cytotoxicity, starting with a maximum applied dose of 1–5 mg/ml for solid test articles or 1–5 μl/ml for liquid test articles and using twofold dilution steps. After an exposure time of 4 hr, the cells should be washed, and a viable cell count obtained the next day. Relative cytotoxicities expressed as the reduction in growth compared to the growth of untreated cells should be used to select 7–10 doses that cover the range from 0 to 50–90% reduction in 24-hr growth. These selected doses should subsequently be applied to cell cultures prepared for mutagenicity testing, but only 4 or 5 of the doses should be carried through the mutant selection process. This procedure compensates for daily variations in cellular cytotoxicity, and ensures the choice of 4 or 5 doses spaced from 9 to 50–90% reduction in cell growth.

Mutagenicity Testing

The procedure used for the nonactivation assay should be based on that reported by Clive and Spector (1975). Cultures exposed to the test article for 4 hr at the preselected doses should be washed and placed in growth medium for 2 or 3 days to allow recovery, growth, and expression of the induced $TK^{-/-}$ phenotype. Cell counts should be determined daily, and appropriate dilutions made to allow optimal growth rates.

At the end of the expression period, 3×10^6 cells for each selected dose should be seeded in soft agar plates with selection medium, and resistant (mutant) colonies counted after 10 days' incubation. To determine the actual number of cells capable for forming colonies, a portion of the cell suspension should also be cloned in normal medium (nonselective). The ration of resistant colonies to total vial cell number will be the mutant frequency.

A detailed flow diagram for the mutation assay is provided in Figure 9.2.

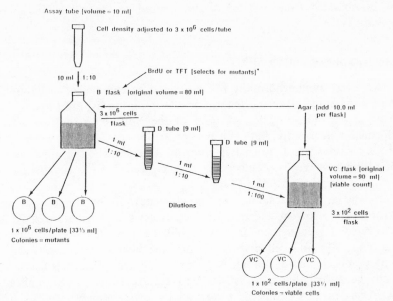

FIGURE 9.2. Lymphoma-cloning flow chart. * Added after removal of 1 ml for viable count dilutions.

The activation assay can be run concurrently with the nonactivation assay. The only difference is the addition of the S9 fraction of rat liver homogenate and necessary cofactors (CORE) during the 4-hr treatment period. CORE consists of NADP (sodium salt) and isocitric acid. The final concentrations of the activation system components in the cell suspension should be 2.4 mg NADP (sodium salt)/ml, 4.5 mg isocitric acid/ml, and 50 μl S9/ml.

Preparation of 9000g Supernatant (S9)

Fischer 344 male rats should normally be used as the source of hepatic microsomes. Induction with Aroclor 1254 or another agent should be performed by injection 5 days prior to kill. After decapitation and bleeding, the liver should be immediately dissected from the animal using aseptic technique and placed in ice-cold 0.25 M sucrose buffered with Tris at pH 7.4. When an adequate number of livers are obtained, wash the collection twice with fresh buffered sucrose and homogenize completely. The homogenate should be centrifuged for 10 min at 9000g in a refrigerated centrifuge, and the supernatant (S9) from this centrifuged sample retained and frozen at $-80°C$ until used in the activation system. The S9 fraction may be obtained from induced or noninduced rats or other species (see Appendix A).

Assay Acceptance Criteria

An assay should normally be considered acceptable for evaluation to test results only if all of the criteria on the following list are satisfied. The activation and nonactivation portions of the mutation assays are usually performed concurrently, but each portion is, in fact, an independent assay with its own positive and negative controls. The activation or nonactivation assays should be repeated independently, as needed, to satisfy the acceptance and evaluation criteria.

1. The average absolute cloning efficiency of the negative controls (average of the solvent and untreated controls) should be between 70 and 130%. A value greater than 100% is possible because of errors in cell counts (usually ±10%) and cell division during unavoidable delays between the counting and cloning of many cell cultures. Cloning efficiencies below 70% do not necessarily indicate substandard culture conditions or unhealthy cells. Assay variables can lead to artificially low cloning efficiencies in the range of 50–70% and still yield internally consistent and valid results. Assays with cloning efficiencies in this range can be conditionally acceptable and dependent on the scientific judgment of the investigator. All assays below 50% cloning efficiency are unacceptable.

2. The solvent and untreated negative controls normally have the same growth rates and cloning efficiencies within experimental error. An unusual effect by the solvent, therefore, indicates an abnormal cell state or an excessive amount of solvent in the growth medium. An assay should be considered unacceptable if the average percent relative growth of solvent controls is less than about 70% of the untreated control value.

3. The average negative control suspension growth factor should not be less than about 15. The optimal value is 25, which corresponds to fivefold increases in cell number for each of the 2 days following treatment of the experimental cultures.

4. The background mutant frequency (average of the solvent and untreated negative controls) is calculated separately for concurrent activation and nonactivation assays, even though the same population of cells is used for each assay. The activation negative controls contain the S9 activation mix and typically have a somewhat higher mutant frequency than the nonactivation negative controls. For both conditions, the normal range of background frequencies for assays performed with different cell stocks is 5 $\times$ 10^{-6} to 50 $\times$ 10^{-6}. Assays with backgrounds outside this range are not necessarily invalid, but should not be used as primary evidence for evaluation of a test article. These assays can provide supporting evidence.

5. A positive control is included with each assay to provide confidence in the procedures used to detect mutagenic activity. The normal range of mutant frequencies induced by 0.5 μl/ml EMS (nonactivation assay) is 300–800 $\times$ 10^{-6}; for 0.3 μl/ml DMN (activation assay), the normal range is 200–800 $\times$ 10^{-6}. (The concurrent background frequencies are subtracted from these values.) These ranges are broad primarily because the effective treatment with these agents is variable between assays. An assay should be considered acceptable in the absence of a positive control (loss due to contamination or technical error) only if the test article clearly shows mutagenic activity as described in the evaluation criteria. If the test article appears to have no, or only weak, mutagenic activity, an acceptable assay must have a positive control mutant frequency above the lower limits of the normal range. Assays in which the normal range is exceeded may require further interpretation.

6. For test articles with little or no mutagenic activity, an assay must include applied concentrations that reduce the suspension growth to 5–10% of the average solvent control or reach the maximum applied concentrations given in the evaluation criteria. Suspension growth is a combined measure of cell death and reduced growth rates. A 5% relative suspension growth, therefore, could correspond to 90% killing followed by growth of the survivors at one-half the normal rate for 1 day and normal growth for the second day. At the other extreme, this condition could be obtained by no killing and complete inhibition of growth for 2 days. a reasonable limit to testing for the presence of mutagenic action is about 80–90% killing of cells. Because of the uncertainty in the actual lethality of treatment in the assay and the fact that mutant frequencies increase as a function of lethality, an acceptable assay for the lack of mutagenic activity must extend to the 5–10% relative suspension growth range. There is no maximum toxicity requirement for test articles that clearly show mutagenic activity.

7. An experimental treatment that results in fewer than 2.5 $\times$ 10$_6$ cells by the end of the 2-day growth period should not be cloned for mutant analysis.

8. An experimental mutant frequency should be considered acceptable for evaluation only if the relative cloning efficiency is 10% or greater and the total number of viable clones exceeds about 20. These limits avoid problems with the statistical distribution of scorable colonies among dishes and allow factors no larger than 10 in the adjustment of the observed number of mutant clones to a unit number of cells (10^6) able to form colonies.

9. Mutant frequencies are normally derived from sets of three dishes for both the mutant colony count and viable colony count. To allow for contamination losses, an acceptable mutant frequency can be calculated from a minimum of two dishes per set if the colony numbers in the two dishes differ by no more than about threefold.

10. The mutant frequencies for five treated cultures should normally be determined in each assay. A required number of different concentrations cannot be explicitly stated, although a minimum of three analyzed cultures is considered necessary under the most favorable test conditions to accept a single assay for evaluation of the test article.

Assay Evaluation Criteria

Mutation assays are initiated by exposing cell cultures to a range of concentrations of test article that is expected, on the basis of preliminary toxicity studies, to span the cellular responses of no observed toxicity to growth to complete lethality within 24 hr of treatment. Then, five dose levels are usally selected for completion of the mutation assay. The doses are selected to cover a range of toxicities to growth with emphasis on the most toxic doses. An assay may need to be repeated with different concentrations to properly evaluate a test article.

The minimum condition considered necessary to demonstrate mutagenesis for any given treatment should be a mutant frequency that exceeds 150% of the concurrent background frequency by at least 10×10^{-6}. The background frequency is defined as the average mutant frequency of the solvent and untreated negative controls. The minimum increase is based on extensive experience which indicates that assay variability increases with higher backgrounds, and this calculated minimum increase is often a repeatable result. Statistical analysis for the confidence limits is not yet available.

The observation of a mutant frequency that meets the minimum criterion for a single treated culture within a range of assayed concentrations is not sufficient evidence to evaluate a test article as a mutagen. The following test results must be obtained to reach this conclusion for either activation or nonactivation conditions.

1. A dose-related or toxicity-related increase in mutant frequency should be observed. It is desirable to obtain this relation for at least three doses, but this depends on the concentration steps chosen for the assay and the toxicity at which mutagenic activity appears.

2. An increase in mutant frequency may be followed by only small or no further increases at higher concentrations or toxicities. However, a decrease in mutant frequency to values below the minimum criterion is not acceptable in a single assay to classify the test article as a mutagen. If the mutagenic activity at lower concentrations or toxicities is large, a repeat assay should be performed to confirm the mutagenic activity.

3. If an increase of about two times the minimum criterion or greater is observed for a single dose near the highest testable toxicity, as defined in Assay Acceptance Criteria, the test article should be considered mutagenic.

Smaller increases at a single dose near the highest testable toxicity require confirmation by a repeat assay.

4. For some test articles the correlation between toxicity and applied concentration is poor. The proportion of the applied article that effectively interacts with the cells to cause genetic alterations is not always repeatable or under control. Conversely, measurable changes in the frequency of induced mutants may occur with concentration changes that cause only small changes in observable toxicity. Therefore, either parameter, applied concentration or toxicity (percent relative growth), can be used to establish whether the mutagenic activity is related to an increase in effective treatment. A negative correlation with dose is acceptable only if a positive correlation with toxicity exists. An apparent increase in mutagenic activity as a function of decreasing toxicity is not acceptable evidence for mutagenicity.

A test article should be evaluated as nonmutagenic in a single assay only if the minimum increase in mutant frequency is not observed for a range of applied concentrations that extends to toxicity causing 5–10% relative suspension growth. If the test article is relatively nontoxic, the maximum applied concentrations should normally be 10 mg/ml (or 10 μl/ml) for articles in organic solvents. If a repeat assay does not confirm an earlier, minimal response as discussed above, the test article should be evaluated as nonmutagenic in this assay system.

Reference

Clive, D., and Spector, J. F. S.: Laboratory procedure for assessing specific locus mutations at the TK locus in cultured L5178Y mouse lymphoma cells. *Mutat. Res.* **31**:17–29, 1975.

PROTOCOL 6

CHROMOSOME ABERRATIONS IN CHINESE HAMSTER OVARY CELLS

The objective of this *in vitro* assay is to evaluate the ability of a test article to induce chromosome aberrations in CHO cells.

Materials

Indicator Cells

Cells to be used in this assay can be obtained from the American Type Culture Collection Repository No. CCL61, Rockville, MD. The original cells were obtained from an ovarian biopsy of a Chinese hamster. This is a permanent cell line with an average cycle time of 10–12 hr.

Medium

CHO cells for this assay should be grown in Ham's F12 medium supplemented with 10% fetal calf serum (FCS). The cells should be split back to 3×10^5 per 75-cm² plastic flask and fed 24 hr prior to treatment with 10 ml of fresh medium.

Control Articles

The solvent for the test article should be used as the solvent or vehicle control article.

EMS, a known mutagen and clastogenic agent, may be dissolved in culture medium and used as a positive control article for the nonactivation studies at a final concentration of 0.5 μl/ml.

CP, a clastogen that requires metabolic transformation by microsomal enzymes, may be used as a positive control article for activation studies at a final concentration of 2×10^{-4} M to 10^{-3} M.

Experimental Design

Toxicity and Dose Determination

The solubility, toxicity, and doses for the test article may be determined prior to screening. The effect of each test article on the survival of the indicator cells should be determined by exposing the cells to a wide range of article concentrations in complete growth medium. Toxicity should be measured as the loss in growth potential of the cells induced by 4-hr exposure to the test article followed by a 24-hr expression period in growth medium. Doses should be selected from the range of concentrations by bracketing the highest dose that shows no loss in growth potentials with at least one higher and three lower doses. Otherwise, a half-log series of doses should be employed, with the highest dose being perhaps limited by solubility, but in any case not to exceed 5 mg/ml. The doses should cover at least four orders of magnitude, and all doses that yield sufficient numbers of scorable metaphase cells considered in the analysis.

Cell Treatment

For the nonactivation assay, approximately 10^6 cells should be treated with the test article at predetermined doses at 37°C in growth medium. Two replicate cultures per dose should be employed in this assay. Incubation should then be continued for 10 hr (M1 cells). Colcemid should be added for the last 3 hr of incubation (2×10^{-7} M final concentration), and metaphase cells collected by mitotic shake-off (Terasima and Tolmach, 1961). These cells should be swollen with 0.075 M KCl hypotonic solution, washed three times in fixation (methanol : acetic acid, 3 : 1), dropped onto slides, and air-dried.

For the activation assay, the test article should be tested in the presence of an S9 rat liver activation system. This assay differs from the nonactivation assay in that S9 reaction mixture should be added to the growth medium, together with the test article, for 2 hr. The exposure period should be terminated by washing the cells twice with saline. From this point they should be treated as described for the nonactivation assay.

Preparation of S9 Reaction Mixture

Fischer 344 male rats normally should be used as the source of hepatic microsomes. Induction with Aroclor 1254 or another agent should be performed by injections 5 days prior to kill. After decapitation and bleeding, the liver should be immediately dissected from the animal using aseptic technique and placed in ice-cold 0.25 M sucrose buffered with Tris at pH 7.4. When an adequate number of livers are obtained, the collection should be washed twice with fresh buffered sucrose and completely homogenized. The homogenate should be centrifuged for 10 min at 9000g in a refrigerated centrifuge, and the supernatant (S9) from this centrifuged sample retained and frozen at $-80°C$ until used in the activation system (see Appendix A). The S9 fraction may be obtained from induced or noninduced rats or other species, as requested. This S9 fraction should be added to a "core" reaction mixture to form the activation system that follows:

Component	Final concentration/ml
NADP (sodium salt)	2.4 mg
Isocitric acid	4.5 mg
Homogenate fraction	15 μl

Staining and Scoring of Slides

Slides should be stained with 10% Giemsa at pH 6.8 for subsequent scoring of chromosome aberration frequencies. Fifty cells should be scored from each of two replicate cultures per dose.

Score and record gaps, breaks, fragments, and reunion figures, as well as numerical aberrations such as polyploid cells, on standard forms. The complete list of aberrations to be scored is:

chromatid gap   pulverized chromosome
chromatid break   pulverized chromosomes
chromosome gap   pulverized cells
chromosome break   complex rearrangement
chromatic deletion   ring chromosome
fragment   dicentric chromosome
acentric fragment   minute chromosome
triradial   greater than 10 aberrations
quadriradial   endoreduplication

For control of bias, all slides should be coded prior to scoring and scored blind.

Evaluation Criteria

A number of general guidelines have been established to aid in determining the meaning of CHO chromosomal aberrations. Basically, an attempt should be made to establish whether a test article or its metabolites can interact with chromosomes to produce gross lesions or changes in chromosome numbers, and whether these are of a type which can survive more than one mitotic cycle of the cell. All aberration figures detected by this assay result from breaks in the chromatin that either fail to repair or repair in atypical combinations.

It is anticipated that many of the cells bearing breaks or reunion figures would be eliminated (i.e., fail to divide again) after their first mitotic division and, as a corollary, that those cells which survive the first division would primarily bear balanced lesions. The detection of these lesions and, hence, a complete risk evaluation, must usually rely on additional testing. In general, a cell bearing configurations such as small deletions or reciprocal translocations may be perpetuated and, therefore, constitute a greater risk to an individual than large deletions or complex rearrangements.

Data should be summarized in tabular form and evaluated. Gaps should not be counted as significant aberrations unless they are present at much higher than usual frequency. Open breaks should be considered indicators of genetic damage, as should configurations resulting from the repair of breaks. The latter includes translocations, multiradials, rings, multicentrics, etc. Reunion figures such as these should be weighted higher than breaks, since they usually result from more than one break and may lead to stable configurations.

The number of aberrations per cell should also be considered significant; cells with more than one aberration indicate more genetic damage than those containing evidence of single events.

Frequently, one is unable to locate 100 suitable metaphase spreads. Possible causes for this appear to be related to cytotoxic effects that alter the duration of the cell cycle, kill the cell, or cause clumping of the chromosomes.

Comparison with a concurrent negative control that happens to show an unusually low frequency of aberrations can suggest undue statistical significance; therefore, treatment data should also be considered against historical control data.

In either event the type of aberration, its frequency, and its correlation to dose trends in a given time period should all be considered in evaluating a test article as being mutagenically positive or negative.

Statistical analysis should employ a two-tailed t-test. This test can be performed on the number of breaks per chromosome in treated and control samples. Dose regression analysis is also useful.

Reference

Terasima, T., and Tolmach, L. J.: Changes in X-ray sensitivity of HeLa cells during the division cycle. *Nature (London)* **190**:1210–1211, 1961.

PROTOCOL 7

SISTER CHROMATID EXCHANGE IN HUMAN LYMPHOCYTES

The objective of this assay is to evaluate the ability of a test article to induce SCE in cultured human lymphocytes (i.e., *in vitro*).

Materials

Human blood should be drawn aseptically into a sterile heparinized syringe and diluted by 30% with acid–citrate–dextrose (ACD) (e.g., 10 ml blood plus 4 ml ACD). This mixture can be stored for at least 3 days at room temperature with no significant loss of visibility and no significant changes in background SCE frequency.

In general, cultures should be initiated by adding 0.4 ml of the above mixture to 5 ml of RPMI 1640 with 15% FCS, 1% penicillin–streptomycin solution, and 0.1 ml phytohemagglutinin (PHA) in sterile glass bottles. Specific protocols to be used are discussed in the following paragraphs.

Immediately prior to each assay, the test article should be diluted in an appropriate solvent (water, DMSO, ethanol, or acetone). A series of dilutions should be performed that will yield the desired set of final concentrations (in culture medium) with the final concentration of solvent being less than 5% whenever possible.

The solvent in which the test article is prepared should be used as the solvent or vehicle control article.

Positive control cultures may be treated with 10^{-3} M EMS, a compound that induces SCEs in the absence of metabolic activation.

Experimental Design

Cells must incorporate BrdU during two periods of DNA replication (S phases) so that subsequently observed metaphase cells (M2 cells) can be stained for sister chromatid differentiation (SCD). Studies have demonstrated that good yields of M2 cells are obtained as early as 66 hr and as late as 78 hr after PHA stimultion of normal lymphocytes. Many substances, however, delay cells in their progression through the mitotic cycle, so slightly longer times may be necessary to obtain an adequate yield of M2 cells. This should be determined experimentally.

Doses should be determined from preliminary tests for mitotic delay and mitotic indices determined in cultures exposed to a log series of doses (e.g., 1, 10, 100×) of the test article. Exposure should be continuous for the entire 72-hr period of incubation. The lowest concentration that reduces the mitotic index to 0–1% should then become the highest concentration employed in the subsequent SCE assay.

A test article may be assayed for its ability to induce SCE in unstimulated (GO) lymphocytes (Option 1) or in stimulated (cycling) lymphocytes (Option 2).

Option 1

In this system, blood (plus ACD) should be added to medium without PHA, and the test article added at up to five preselected dose levels (plus positive and negative controls) with two replicates per dose. After 4 hr at 37°C, the cells should be washed twice with phosphate buffered saline (PBS). Fresh medium (plus PHA) should then be added together with BrdU (25 μM final concentration), and cultures maintained at 37°C in complete darkness for an additional 76 hr (more if required). Colcemid (10^{-6} M final concentration) should be added for the last 4 hr to arrest cells in metaphase. Harvesting procedures to be used are described below.

Option 2

In this system, cultures should be initiated with PHA present to stimulate progression through the mitotic cycle. After 24 hr, when cells have begun to cycle, the test article should be added as above, together with BrdU (25 μM final concentration). Cultures then should be maintained at

37°C in the dark for an additional 52 hr. Colcemid should be present during the last 4 hr. It is believed that with this option, lymphocytes will be exposed to the test article in all phases of the mitotic cycle. In this way, phase-specific agents (e.g., agents to which DNA is accessible only during S) will give positive results vis-à-vis Option 1.

Lymphocyte harvest in both systems should involve centrifugation of the cell suspension, decanting the supernatant, and a 5- to 10-minute hypotonic treatment with 0.075 M KCl. Then, cells should be washed three times in methanol: acetic acid (3:1) and dropped onto slides to air-dry.

Staining and Scoring of Slides

Staining for SCD should be accomplished with a modified fluorescent plus Giemsa (FPG) technique (after Perry and Wolff, 1974). Slides should be stained for 10 min with Hoechst 33258 (5 μg/ml) in M/15 Sorensen's buffer (pH 6.8), mounted in the same buffer, and exposed to UV light from a mercury lamp for the amount of time required for SCD. Following UV exposure, slides should be stained with 10% Giemsa for 10–20 min and then mounted in Depex.

M2 cells should be scored for the frequency of SCEs per cell and per chromosome. Twenty-five cells should be scored from each of two replicate cultures (50 cells per dose).

For control of bias, all slides should be coded prior to scoring and scored blind.

Evaluation Criteria

Data are generally presented in table form. Interpretation should be based on the increase in SCE frquency as a function of dose and/or on the statistical significance of increases above the background or "spontaneous" level. Statistical analysis will employ a two-tailed t-test, and an SCE frequency increase should be considered positive if $p < 0.05$.

Reference

Perry, P., and Wolff, S.: New Giemsa method for the differential staining of sister chromatids. *Nature (London)* **251**:156–158, 1974.

PROTOCOL 8

SISTER CHROMATID EXCHANGE IN CHINESE HAMSTER OVARY CELLS

The objective of this *in vitro* assay is to evaluate the ability of a test article to induce SCE in CHO cells, with and without metabolic activation.

Materials

Cells to be used in this assay can be obtained from the American Type Culture Collection Repository No. CCL61, Rockville, MD. The original cells were obtained from an ovarian biopsy of a Chinese hamster. This is a permanent cell line with an average cycle time of 10–12 hr.

CHO cells for this assay should be grown in McCoy's 5A medium supplemented with 10% FCS. The cells should be split back to 3×10^5 per 75-cm^2 plastic flask and fed 24 hr prior to treatment with 10 ml of fresh medium.

The solvent for the test article should be used as the solvent or vehicle control article.

EMS, a clastogen that induces SCE, may be dissolved in culture medium and used as a positive control article for the nonactivation studies at a final concentration of 0.3 μl/ml.

CP an SCE inducer that requires metabolic biotransformation by microsomal enzymes, may be used as the positive control article for activation studies at a final concentration of 5×10^{-6} M to 10^{-5} M.

Experimental Design

Toxicity and Dose Determination

The solubility, toxicity, and doses for the test article may be determined prior to screening. The effect of each test article on the survival of the indicator cells should be determined by exposing the cells to a wide range of article concentrations in complete growth medium. Toxicity should be measured as the loss in growth potential of the cells induced by 4-hr exposure to the test article followed by a 24-hr expression period in growth medium. Doses should be selected from the range of concentrations by bracketing the highest dose that shows no loss in growth potential with at least one higher and three lower doses. Otherwise, a half-log series of doses should be employed, with the highest dose being perhaps limited by solubility, but in any case not to exceed 5 mg/ml. The doses should cover at least four orders of magnitude, and all doses that yield sufficient numbers of scorable metaphase cells should be considered in the analysis.

Cell Treatment

For the nonactivation assay, approximately 10^6 cells should be treated in growth medium with the test article at predetermined doses, and then incubated at 37°C for 2 hr on a rocker. Then, add BrdU (10 μM final concentration) to the culture tubes and continue incubation in the dark for 24–30 hr. Longer times will often be necessary to permit cell passage through two DNA replication cycles in the presence of BrdU following treatments that cause mitotic delay. This can be determined experimentally. Colcemid should be added for the last 3 hr of incubation (final concentration 2×10^{-7} M), and metaphase cells collected by mitotic shake-off (Terasima and Tolmach, 1961). These cells should be swollen with 0.075 M KCl hypotonic solution, then washed three times in fixative (methanol: acetic acid, 3:1), dropped onto slides, and air-dried.

For the activation assay, the test article should be tested in the presence of an S9 rat liver activation system. This assay differs from the nonactivation assay in that S9 reaction mixture should be added to the growth medium, together with the test article, for 2 hr. The exposure period should be terminated by washing the cells twice with saline. From this point they should be treated as described for the nonactivation assay.

Preparation of S9 Reaction Mixture (see Appendix A)

Fischer 344 male rats should normally be used as the source of hepatic microsomes. Induction with Aroclor 1254 or another agent should be performed by injections 5 days prior to kill. After decapitation and bleeding, the liver should be immediately dissected from the animal using aseptic technique and placed in ice-cold 0.25 M sucrose buffered with Tris at pH 7.4. When an adequate number of livers is obtained, the collection should be washed twice with fresh buffered sucrose and completely homogenized. The homogenate should be centrifuged for 10 min at 9000g in a refrigerated centrifuge, and the supernatant (S9) from this centrifuged sample retained and frozen at $-80°$C until used in the activation system. The S9 fraction may be obtained from induced or noninduced rats or other species. This S9 fraction should be added to a "core" reaction mixture to form the activation system that follows:

Component	Final concentration/ml
NADP (sodium salt)	2.4 mg
Isocitric acid	4.5 mg
Homogenate fraction	15 μl

Staining and Scoring of Slides

Slides should be stained for 10 min with Hoechst 33258 (5 μg/ml) in M/15 Sorensen's buffer (pH 6.8), mounted in the same buffer, and exposed to UV light from a mercury lamp for the amount of time required for SCD.* Following UV exposure, the slides should be stained with 10% Giemsa for 10 min and then mounted in Depex.

* Modification of Perry and Wolff (1974) technique.

Second-division cells (M2 cells) should be scored for the frequency of SCEs per cell and per chromosome. The proportions of cells in the first, second, and third divisions (i.e., M1, M2, and M3 cells) should also be determined by scoring 100 cells. For SCE analysis, 50 M2 cells should typically be scored.

For control of bias, all slides should be coded prior to scoring and scored blind.

Evaluation Criteria

Interpretation of data should be based on the increase in SCE frequency as a function of dose and/or on the statistical significance of increases above the background or "spontaneous" level. The t-statistic should be calculated, and an SCE frequency increase considered positive if $p < 0.05$.

References

Perry, P., and Wolff, S.: New Giemsa method for the differential staining of sister chromatids. *Nature (London)* **251**:156–158, 1974.
Terasima, T., and Tolmach, L. J.: Changes in X-ray sensitivity of HeLa cells during the division cycle. *Nature (London)* **190**:1210–1211, 1961.

PROTOCOL 9

DETERMINATION OF UNSCHEDULED DNA SYNTHESIS IN HUMAN WI-38 CELLS

This assay is used to evaluate the potential of a test article to cause DNA damage repairable by new DNA synthesis. The objective is to evaluate a test article for its ability to induce UDS in WI-38 human cells as determined by the amount of radioactive thymidine incorporated per microgram of DNA. Untransformed human cells in culture (WI-38) are used to test the response of human DNA repair mechanisms to DNA-damaging agents.

WI-38 cells attach to a surface and multiply until contact inhibition imposed by a confluent monolayer causes complete growth arrest. Cultures thus arrested are composed primarily of cells which are not actively synthesizing DNA. A further block of DNA synthesis is achieved by inclusion of hydroxyurea in the medium. If tritiated thymidine ([³H]-TdR) is introduced in the medium in the presence of hydroxyurea, little or no label is incorporated into DNA. Addition of a test article that interacts with the DNA usually stimulates a repair response in which the altered portion of

DNA is excised and the missing region replaced by DNA synthesis. This synthesis of DNA by nondividing cells is known as UDS and can be measured by determining the amount of [³H]-TdR incorporated into DNA. Exposure of WI-38 cells to various forms of radiation or to articles known to be mutagenic or carcinogenic has resulted in the observation of UDS.

Materials

The WI-38 normal human diploid cell strain can be obtained directly from the American Type Culture Collection, Rockville, MD, at the lowest passage available. Laboratory cultures should be grown in EMEM supplemented with 10% fetal bovine serum, 2 mM glutamine, and 100 units of penicillin and 100 μg of streptomycin/ml.

A negative control article consisting of assay procedures on untreated cells should be utilized in all cases. If the test article is not soluble in growth medium, an organic solvent (normally DMSO) should be used; the final concentration of solvent in the growth medium should be 1% or less. Cells exposed to solvent in the medium should be assayed as the solvent negative control article to determine any effects on UDS caused by the solvent alone. For test articles assayed with activation, untreated and solvent negative control articles should include the activation mixture.

N-Methyl-N'-nitro-N-nitrosoguanidine (MNNG) is highly mutagenic via alkylation of DNA and may be used at a final concentration of 5–10 μg/ml as a positive control article for assays performed without activation. Benzo[a]pyrene requires metabolic activation to react with DNA and may be used at 1–20 μg/ml as a positive control article for activation conditions.

Solid articles can be dissolved in growth medium, if possible, or in DMSO. Liquids can be tested by direct addition to the test system at predetermined concentrations or following dilution in a suitable solvent.

Experimental Design

Dose Selection

An assay should be performed with at least four doses of the test article, starting generally with 5 mg/ml (or 5 μl/ml for a liquid) and diluting, alternating fivefold and twofold dilution steps, to 0.1 mg/ml (or 0.1 μl/ml) or less. Doses which result in excess toxicity can be recognized by UDS activities less than control values. If necessary, the assay should be repeated with four doses spaced below the excess toxicity level.

Nonactivation Assay

The assay presented here is based on the procedure give by Stich and Laishes (1973). WI-38 cells should be seeded at 2.5×10^5 cells per 100-mm dish and grown to confluency. The serum concentration should then be reduced from 10 to 0.5% for another 5 days to help ensure synchronization of the cells into a state of no DNA synthesis. To avoid a spurt of DNA synthesis when the medium is changed prior to treatment, hydroxyurea should be added to the medium 0.5 hr before the addition of test article. Selected doses of the test article and a positive article should be applied (two dishes per dose), and additional dishes treated with solvent if appropriate. [^{3}H]-TdR should be added to all cultures simultaneously with the test article. After an exposure period of 1.5 hr at 37°C, the dishes should be washed free of test article and label, and stored frozen until DNA extraction can be performed.

Activation Assay

The activation assay can be run separately or concurrently with the nonactivation assay. The only difference would be the addition of an activation mixture to the cells immediately before the test article. The activation mixture should consist of S9 fraction of rat liver homogenate and the cofactor mixture, NADP and isocitric acid (see Appendix A).

Extraction of DNA

The cells from the frozen stored dishes should be removed by scraping in 0.5% sodium lauryl sulfate (SLS) in 0.017 M disodium citrate and placed into centrifuge tubes. The two dishes for each dose of test article should be combined in this procedure. DNA should be precipitated by the addition of at least two volumes of 95% ethanol. The mixture should be centrifuged, and the pellets washed with 95% ethanol and resuspended in absolute ethanol:ethyl ether (2:1). The tubes should be incubated at 70°C for 3 min to solubilize nonpolymeric contaminants, then centrifuged and the pellets washed with 95% ethanol. The pellets should next be dissolved in 0.3 N NaOH and incubated at 70°C for 20 min to hydrolyze any RNA contaminant. The DNA should then be reprecipitated by chilling and the addition of an equal volume of 2 N perchloric acid in an ice bath. After centrifugation and a 0.2 N perchloric acid wash, the DNA pellet should be dissolved in 1 N perchloric acid by heating at 70°C. Any remaining insoluble article should be removed by centrifugation, and the supernatant analyzed for DNA and radiolabel.

DNA Analysis

Samples from each tube should be removed for determination of absorbance at 260, 280, and 320 nm and the micrograms of DNA calculated from the 260-nm absorbance reading based on a standard curve of hydrolyzed DNA. The other absorbance values should be used as indicators of purity (280 nm) and turbidity (320 nm) in the assayed solutions of DNA.

Samples (0.5 ml) from each tube should also be placed into 10 ml of scintillation fluid, and the amount of ^{3}H label counted by a liquid scintillation spectrometer. Convert the counts into disintegrations per minute (dpm).

The results for each sample, expressed as dpm per microgram of DNA, should be compared to the UDS activity of the appropriate negative control.

Preparation of 9000g Supernatant (S9)

Fischer 344 male rats should normally be used as the source of hepatic microsomes. Induction with Aroclor 1254 or other agent should be performed by injections 5 days prior to kill. After decapitation and bleeding, the liver should be immediately dissected from the animal using aseptic techniques and placed into ice-cold 0.25 M sucrose buffered with Tris at pH 7.4. When an adequate number of livers are obtained, the collection should be washed twice with fresh buffered sucrose and completely homogenized. The homogenate should be centrifuged for 10 min at 9000g in a refrigerated centrifuge, and the supernatant (S9) from this centrifuged sample retained and frozen at $-80°$C until used in the activation system. The S9 fraction may also be obtained from noninduced rats or from other species.

Evaluation Criteria

The laboratory procedures, DNA synthetic activity (dpm) per unit amount of DNA, and DNA synthetic activity relative to controls should be provided for four or five applied concentrations of test article.

Several criteria have been established which, if met, provide a basis for declaring an article genetically active in the UDS assay. These criteria are derived from a historical data base and are helpful in maintaining uniformity in evaluations from article to article and run to run. While these criteria are reasonably objective, a certain amount of flexibility may be required in making the final evaluations since absolute criteria may not be applicable to all biological data.

An article should be considered active in the UDS assay if (1) The DNA synthetic activity for a treated culture is at least 150% of the negative control value, and (2) a dose–response relationship is observed for at least two consecutive dose levels.

All evaluations of UDS activity should be based on the concurrent solvent and positive control values for each assay. Replicate assays will improve the reliability of this procedure and permit a statistical analysis of the data.

Reference

Stich, H. F., and Laishes, B. A.: DNA repair and chemical carcinogens. *Pathobiol. Ann.* 3:341–376, 1973.

PROTOCOL 10

UNSCHEDULED DNA SYNTHESIS
IN RAT LIVER PRIMARY CELL CULTURES

This assay is a highly sensitive test designed to measure UDS in primary rat liver cell (hepatocyte) cultures using the autoradiographic technique described by Williams (1977). Primary hepatocytes have the advantage of having sufficient metabolic activity to eliminate the need for the addition of a microsomal activation system.

The existence and degree of DNA damage can be inferred from an increase in nuclear grain counts compared to untreated hepatocytes. The types of detectable DNA damage are unspecified, but are recognizable by the cellular repair system and result in the incorporation of new bases (including [^{3}H]-TdR) into the DNA.

Materials

The indicator cells for this assay are hepatocytes obtained from adult male Fischer 344 rats (weighing 150–300 g). The animals can receive a commercial diet and water *ad libitum* prior to use.

The cells should be cultured in Williams' Medium E (WME) supplemented with 10% fetal bovine serum, 2 mM L-glutamine, and 125 μg/ml gentamycin. (This medium is referred to as complete WME; incomplete WME contains no serum.)

A negative control consisting of assay procedures performed on untreated cells should be performed in all cases. If the test article is not soluble in water, a stock solution in an organic solvent (normally DMSO) should be prepared; the final concentration of solvent in the growth medium should be 1% or less in the treated cultures and the negative control.

The positive control article should be known to induce UDS in rat hepatocyte primary cell cultures, such as 2-AAF at 2×10^{-3} M (400 μg/ml). Aflatoxin B$_1$ (AFB$_1$) at 2×10^{-4} M (60 μg/ml) may also be added or substituted for 2-AAF.

Experimental Design

All cultures should be started at the same time. Some of them should be used for a preliminary cytotoxicity test for dose selection; the remaining cultures should be incubated overnight at 37°C for the UDS assay.

Dose Selection

The preliminary cytotoxicity test should be initiated with a series of applied concentrations of the test article, starting at a maximum concentration of 5000 μg/ml (or 5 μl/ml) and diluting in twofold steps to about 0.6 μg/ml (0.6 μl/ml). Cells should be exposed for a period of 1 hr approximately 2 hr after initiation of the primary cultures. After removal of the test article, cells should be incubated an additional 2 hr in WME. A viable cell count using trypan blue exclusion should be obtained; and those treatments that reduce the number of viable cells below about 50%, relative to the negative control, should be eliminated from further testing. A second viable cell count should be obtained for the remaining treated cultures at about 24 hr. At least five doses which span the range from *no apparent toxicity* to *complete loss of viable cells* in about 24 hr should be chosen for the UDS assay.

UDS Assay Method

The assay described here is based on the procedures described by Williams (1977). The hepatocytes should be obtained by perfusion of livers *in situ* for 4 min with Hanks' balanced salts (Ca^{2+}- and Mg^{2+}-free) containing 0.5 mM ethyleneglycol-bis(β-aminoethyl ether)-N,N-tetraacetic acid (EGTA) and HEPES buffer at pH 7.0. Then incomplete WME with 100 units/ml of Type I collagenase should be perfused through the liver for 10 min. The hepatocytes should be dispersed by scraping the excised livers with a sterile steel comb in a culture dish containing incomplete WME and collagenase. After centrifugation to remove the collagenase, the cells should be resuspended in complete WME and counted. A series of 35-mm culture dishes, each containing a 25-mm round plastic coverslip, should be inoculated with approximately 0.5×10^6 viable cells in 3 ml complete WME per dish.

An attachment period of 1.5 hr at 37°C in a humidified atmosphere containing 5% CO_2 should be allowed. Unattached cells should then be removed, and the cultures refed with 2.5 ml complete WME. Some of the cultures can be used for the preliminary cytotoxicity test, while the remaining cultures should be incubated at 37°C until the next day for the UDS assay.

The UDS assay should be initiated by replacing the media in the culture dishes with 2.5 ml WME containing only 0.1% fetal bovine serum and the test article at the desired concentration. If the test article is dissolved in DMSO, 25-μl aliquots of appropriate stock solutions should be added to 2.5 ml of media (0.1% serum) in the culture dishes. Each treatment, including the positive and negative controls, should be performed on six cultures. After treatment for 1 hr, the test article is removed, and the cell monolayers washed twice with incomplete WME. Three of the cultures for each treatment should be used to monitor the toxicity of treatment; these cultures should be refed with complete WME and returned to the incubator. The other three cultures from each treatment should be refed with 2.5 ml complete WME containing 1 μCi/ml of [^{3}H]-TdR and incubated for 3 hr. The labeling will be terminated by washing the cultures with complete WME containing 1 mM thymidine. The toxicity of each treatment should be monitored by performing viable cell counts on one culture 2 hr after treatment, and on two cultures about 24 hr later.

The nuclei in the labeled cells should be swollen by placement of the coverslips in 1% sodium citrate for 10 min. Then the cells should be fixed in acetic acid:ethanol (1:3) and dried for at least 3 days. The coverslips should be mounted on glass slides (cells up), dipped in Kodak NTB2 emulsion, and dried. The coated slides should be stored for 2 weeks at 4°C in light-tight boxes containing packets of Drierite. The emulsions should then be developed in D19, fixed, and stained with Williams' modified hematoxylin and eosin.

The cells should be examined microscopically at approximately 1500× maginification under oil immersion, and the field displayed on the video screen of an automatic counter. UDS should be measured by counting nuclear grains and subtracting the average number of grains in three nuclear-sized areas adjacent to each nucleus (background count). This value is referred to as the net nuclear grain count. The coverslips should be coded to prevent bias in grain counting.

Evaluation Criteria

The net nuclear grain count should be determined for 50 randomly selected cells on each coverslip, whether or not the nuclei contain grains. Only nuclei appearing normal should be scored, and any occasional nuclei blackened by grains too numerous to count should be excluded as cells in which replicative DNA synthesis occurred rather than repair synthesis. If the actual count for any nucleus is less than zero (i.e., cytoplasmic count is greater than nuclear count), a net value of zero should be used in the calculation of the mean value. The mean net nuclear grain count should be determined from the triplicate coverslips (150 total nuclei) for each treatment condition.

Several criteria have been established which, if met, will provide a basis for evaluation of a test article as active in the UDS assay. These criteria are formulated on the basis of published results and laboratory experience and can be used in lieu of a statistical treatment to indicate a positive response. While the criteria are arbitrary guidelines that may not be applicable to all assays and may need revision as the data base increases, they represent a reasonable approach to the evaluation of a test article.

The test article should be considered active in the UDS assay at applied concentrations that cause (1) An increase in the mean nuclear grain count to at least six grains per nucleus in excess of the concurrent negative control value; and/or (2) the percentage of nuclei with six or more grains to increase above 10% of the examined population, in excess of the concurrent negative control; and/or (3) the percentage of nuclei with 20 or more grains to reach or exceed 2% of the examined population.

Generally if the first condition is satisfied, the second and often the third condition will also be met. However, satisfaction of only the second or third condition can also indicate UDS activity. Different DNA-damaging agents can give a variety of nucler labeling patterns, and weak agents may strongly affect only a small minority of the cells. Therefore, all three of the above conditions should be considered in an evaluation. If the negative control shows an average of six grains per nucleus, or 1% of the cells have 20 grains per nucleus, the assay should normally be considered invalid.

A dose-related increase in UDS for at least two consecutive applied concentrations is also desirable to evaluated a test article as active in this assay. In some cases, UDS can increase with dose and then decrease to near zero with successively higher doses. If this behavior is associated with increased toxicity, the test article can be evaluated as active. If an isolated increase occurs for a treatment far removed from the toxic doses, the UDS should be considered spurious.

The test article should be considered inactive in this assay if none of the above conditions is met, and if the assay includes the maximum applied dose or other doses that are shown to be toxic by the survival measurements. If no toxicity is demonstrated for doses below the maximum applied dose, the assay should be considered inconclusive and repeated with higher doses.

The positive control values should not be used as reference points to measure the UDS activity of the test article. UDS elicited by test agents in this assay is probably more dependent on the type of DNA damage inflicted and the available repair mechanisms than on the potency of the test agent as a mutagen or carcinogen. Some forms of DNA damage are repaired without the incorporation of new nucleic acids. Thus, the positive controls should only be used to demonstrate that the cell population employed was responsive and the methodology adequate for the detection of UDS.

Reference

Williams, G. M.: Detection of chemical carcinogens by unscheduled DNA synthesis in rat liver primary cell culture. *Cancer Res.* **37**:1845–1851, 1977.

PROTOCOL 11

IN VITRO TRANSFORMATION OF BALB/3T3 CELLS

This assay evaluates the carcinogenic potential of a test article for its ability to induce transformed BALB/3T3 cells using cultured mammalian cells. The assay is semiquantitative and is used to establish the malignant tranforming activities of mutagens and other test articles on mammalian cells *in vitro*. Approximately 500 mg (or equivalent) of the test article is needed to perform the study.

BALB/3T3 mouse cells will multiply in culture until a monolayer is achieved and will then cease further division. These cells, if injected into immunosuppressed, syngeneic host animals, will not produce neoplastic tumors. However, cells treated *in vitro* with chemical carcinogens will give rise to foci of cellular growth superimposed on the cell monolayer. If these foci are picked from the cultures, grown to larger numbers, and injected into animals, a malignant tumor will in most cases be obtained. Thus, the appearance of piled-up colonies in treated cell cultures, but not in control cultures, is highly correlated with malignant transformation.

Materials

Subclones of Clone I_{13} of BALB/3T3 mouse cells, selected for low spontaneous frequencies of foci formation, should be used for assays. Stocks should be maintained in liquid nitrogen, and laboratory cultures checked periodically to ensure the absence of mycoplasma contamination. Supplemented with 10% fetal bovine serum, cultures should be grown and passaged weekly in EMEM.

A negative control consisting of assay procedures performed on untreated cells should be performed in all cases. If the test article is not soluble in growth medium, and organic solvent (normally DMSO will be used; the final concentration of solvent in the growth medium should be 1% or less. Cells exposed to solvent in the medium should also be assayed as the solvent negative control to determine any effects on survival or transformation caused by the solvent alone. Fifteen flasks of the appropriate type of negative control should be prepared for each assay.

A known carcinogen, such as MCA may be used as a positive control for the transformation of 3T3 cells. Fifteen flasks may be treated with 5 μg MCA/ml for each assay.

All types of articles can be evaluated in the transformation assay. Solids should be dissolved in growth medium, if possible, or in DMSO. Liquids should be tested by direct addition to the test system at predetermined concentrations or following dilution in a suitable solvent. Highly volatile liquids which must be tested in the vapor phase can be added to an airtight container of fixed volume and allowed to completely volatilize in the presence of the exposed cell population. Gases should be treated by measuring known volumes of gas into an airtight container of fixed volume.

Experimental Design

Dose Selection

The solubility of the test article in growth medium, DMSO, or other solvent should be determined first. Fifteen dose levels of the test article should then be chosen, starting with a maximum applied dose of 1 mg/ml for solids or 1 μl/ml for liquid samples and decreasing in twofold-dilution steps. Each dose should be applied to three culture dishes seeded 24 hr earlier with 200 cells per dish.

After an exposure period of 3 days, the cells should be washed and incubated in growth medium for an additional 4 days. The surviving colonies should be stained and counted by hand tally or an automatic colony counter. A relative survival for each dose should be obtained by comparing the number of colonies surviving treatment to the colony counts in negative control dishes. The highest dose chosen for subsequent transformation assays should cause no more than a 50% reduction in colony forming ability. Four lower doses (including one or two doses with no apparent toxicity) should also be selected for the transformation assay.

Transformation Assay

The procedure given here is based on that reported by Kakunaga (1973). Twenty-four hours prior to treatment, a series of 25-cm^2 flasks should be seeded with 10^4 cells per flask and incubated. Fifteen flasks should then be treated for each of the following conditions: five preselected doses of test article; positive control; and negative control. Incubated the flasks for a 3-day exposure period; wash cells and continue incubation for 4 weeks with refeeding twice a week. The assay should be terminated by fixing the cell monolayers with methanol and staining with Giemsa. The stained flasks should be examined by eye and microscope to determine the number of foci of transformed cells.

Scoring of Transformed Foci

Flasks should be coded to prevent bias in the scoring of foci.

At the end of the 4-week incubation period, cultures of normal cells will yield a uniformly stained monolayer of round, closely packed cells. Transformed cells form a dense mass (focus or colony) that stains deeply (usually blue) and is superimposed on the surrounding monolayer of normal cells. The foci are variable in size.

Scored foci have several variations in morphologic features. Most foci consist of a dense piling up of cells and exhibit a random, criss-cross orientation of fibroblastic cells at the periphery of the focus. Other scored foci are composed of more rounded cells with little crisscrossing at the periphery, but with necrosis at the center caused by the dense piling up of a large number of cells. A third variation is a focus without the necrotic center and large number of cells, but which exhibits the crisscross pattern of overlapping cells throughout most of the colony.

Some foci should not be scored. These include small foci of transformed morphology that are found in close proximity to larger foci; these foci should be regarded as being formed from cells which migrated from the larger colony. Other foci which should not be scored will be small areas where some piling up of rounded cells has occurred, but the random orientation of fibroblastic cells is not observed. Microscopic examination should be employed for the final judgment of transformed character for any marginal foci.

Most transformed clones will produce malignant tumors when collected from an unstained transformation plate and injected into syngeneic host animals. Although not routinely performed, this confirmation step can be conducted.

Assay Acceptance Criteria

The assay should be considered acceptable for evaluation if the following criteria are met:

1. The negative control flasks consist of a contiguous monolayer of cells which may or may not contain transformed foci. The lack of contiguous sheet of cells indicates growth conditions too poor to allow reliable detection of weak tranforming agents.
2. The negative control transformation frequency does not exceed an average of about two foci per flask. Attempts should be made to isolate and maintain cell stocks (subclones of BALB/3T3 I_{13}) with a very low spontaneous frequency of transformation.
3. The positive control yields an average number of foci per flask that is significantly different from the negative control at the 99% confidence level (Kastenbaum and Bowman, 1970).

4. A minimum of eight flasks per test condition are available for analysis. At least four dose levels of test article should be assayed.
5. The dose range of test article assayed falls within the 50–100% survival range as determined by the preliminary toxicity test, which measures relative cloning efficiencies.

Assay Evaluation Criteria

In many cases, no transformed foci will be observed in the set of flasks composing the negative control. This does not necessarily mean that any foci found in the treated flasks constitute a positive response in this assay. To determine what minimum number of foci will allow a conclusion that the frequency of transformed foci has been elevated over the negative control, a historical negative control data base should be used. This data base should consist of the 10 most recent assays in which 100–150 negative control flasks have been scored.

The statistical tables provided by Kastenbaum and Bowman (1970) should be used to determine whether the results at each dose level are significantly different from the historical control at the 95% or 99% confidence levels. This test compares variables distributed according to Poissonian expectations by summing the probabilities in the tails of two binominal distributions. The 95% confidence level must be met to consider the test article active at a particular dose level.

If the negative control is found by the same test to be significantly different from the historical control ($p \leq 0.05$), the assay should be elevated independently. Comparisons between the current negative control and tested dose levels should be analyzed by the Kastenbaum–Bowman tables.

The numbers of induced foci usually does not increase proportionately with the applied dose in this assay. In fact, above a minimum dose level where the number of foci is elevated, further increases in dose may result in little or no increase in the number of foci. The number of foci can be reduced at the highest dose assayed if the toxicity is too high. A response at only one dose level (other than the highest tested dose) that just meets the 95% confidence level should normally not be considered sufficient evidence for activity in this assay. All other degrees of response should provide evidence for classifying a test article as active, although one may have to exercise scientific judgment and/or obtain expert opinion in the evaluation.

References

Kakunaga, T.: A quantitative system for assay of malignant transformation by chemical carcinogens using a clone derived from BALB/3T3. *Int. J. Cancer* **12**:463–473, 1973.
Kastenbaum, M. A., and Bowman, K. O.: Tables for determining the statistical significance of mutation frequencies. *Mutation Res.* **9**:527–549, 1970.

PROTOCOL 12

MICROBIAL HOST-MEDIATED ASSAY

The objective of this assay is to evaluate the genotoxic activity of a test article on microbial cells implanted in the host animal (Gabridge and Legator, 1969). The HMA employs microbial indicator cells to assay for mutagens using the intact animal as the source of chemical biotransformation, detoxification, systemic distribution, and excretion.

Microbial cells should be injected intravenously (IV) into mice which have been previously exposed for 5 days to various concentrations of the test article (Mohn *et al.*, 1975). Following 3 hr incubation, kill the animals and aseptically remove the microbial cells from the liver.* The cells removed from individual animals should be assayed for mutation induction using conventional microbial plating methods.

Materials

Randomly bred male mice, strain CD-1, can be purchased from Charles River Breeding Laboratories, Inc., Wilmington, MA, or Portage, MI.

The *S. typhimurium* strains to be used in this assay can be obtained from Dr. Bruce Ames, University of California at Berkeley (Ames *et al.*, 1972; Ames *et al.*, 1973a,b; McCann *et al.*, 1975a,b). The strains shown in the following table should be used for this study.

| Strain designation | Gene affected | Additional mutations | | | Mutation type detected |
		Repair	LPS	R Factor	
TA-1535	*hisG*	$\Delta uvrB$	*rfa*	—	Base-pair substitution
TA-1537	*hisC*	$\Delta uvrB$	*rfa*	—	Frameshift
TA-98	*hisD*	$\Delta uvrB$	*rfa*	pKM101	Frameshift
TA-100	*hisG*	$\Delta uvrB$	*rfa*	pKM101	Base-pair substitution

All of these strains have mutations in the histidine operon, mutation (rfa^-), that leads to defective lipopolysaccharide coat, a deletion that covers genes involved in the synthesis of vitamin biotin (bio^-) and in the repair of UV-induced DNA damage ($uvrB^-$). The rfa^- mutation makes the strains more permeable to many large molecules. The $uvrB^-$ mutation decreases repair of some types of chemically or physically damaged DNA and thereby

---

* Cells can also be recovered from other organs such as the lungs, kidneys, spleen, or testes.

enhances the strain's sensitivity to some mutagenic agents. The resistant transfer factor plasmid (R factor) pKM101 in TA-98 and TA-100 is believed to cause an increase in error-prone DNA repair that leads to many more mutations for a given dose of most mutagens (McCann et al., 1975a). In addition, plasmid pKM101 confers resistance to the antibiotic ampicillin, which is a convenient marker to detect the presence of plasmid in the cells.

All indicator strains should be kept at 4°C on minimal medium plates supplemented with a trace of biotin and an excess of histidine. The plates with plasmid-carrying strains should also contain ampicillin (25 μg/ml) to ensure stable maintenance of plasmid pKM101. New stock culture plates should be made every 2 months from the frozen master cultures or from single-colony reisolates that have been checked for their genotypic characteristics (his, rfa, urvB, bio) and for the presence of plasmid. For each experiment, an inoculum from the stock culture plates should be grown overnight at 37°C in nutrient broth (Oxoid CM67) and used.

The bacterial strains should be cultured in Oxoid Media #2 (Nutrient Broth). The selective medium should be Vogel Bonner Medium E with 2% glucose (Vogel and Bonner, 1956). The overlay agar, when employed, should consist of 0.6% purified agar with 0.5 mM histidine, 0.05 mM biotin, and 0.1 M NaCl according to the methods of Ames et al. (1975).

Experimental Design

Animal Husbandry

Animals should be group housed in shoebox cages on AB-SORB-DRI bedding with 10 mice per cage and quarantined for at least 1 week before being placed on study. A commercial diet and water can be made available ad libitum unless contraindicated by the particular experimental design.

For control of bias, animals should be assigned to study groups at random. Prior to study initiation, animals should be weighed, and an overall mean weight determined. The volume of test article administered per animal should be established using this method unless there is significant variation among individuals, in which case individual calculations should be made. Animals should be uniquely identified by either ear tag or ear punch, and treatment groups by cage card.

Sanitary cages and bedding should be used. Personnel handling animals or working within the animal facilities should be required to wear suitable protective garments. When appropriate, individuals with respiratory or other overt infections should be excluded from the animal facilities.

Dose Selection (see Appendix B)

If acute toxicity information (e.g., LD_{50}) is available, it should be used to determine dose levels as described below. If it is not available, dose levels should be determined using five groups of six animals each in a toxicity study. The LD_{50} should be determined statistically by probit analysis (Finney, 1971). In the event that an LD_{50} cannot be determined because the test article is nontoxic, the doses for the mutation studies should be selected as high in relation to conditions of use.

Normally, if the LD_{50} can be approximated, the highest dose for acute studies should be arbitrarily selected as one-half the LD_{50}. One-third and one-tenth this high dose should be used as the intermediate and low dose levels, respectively.

Route of Administration

Oral gavage should be employed as the route of administration, and the selected concentrations administered daily up to 5 days via this route. In the event that test article characteristics preclude oral gavage, intramuscular (IM) or intraperitoneal (IP) injection should be employed. These routes of administration are the most common for this test procedure.

Test and Control Articles

The test and control articles should be prepared and administered to mice fasted for 16 hr according to the following procedures:

Group	Number of animals	Animals dosed prior to recovery (hr)
Solvent	10	3
Low (test article)	10	3
Intermediate (test article)	10	3
High (test article)	10	3
Positive control[a]	10	3

[a] DMN (50 mg/kg) or 2-AAF (100 mg/kg) may be used as the positive control agent and should be administered either IM or IP.

Bacteria Administration and Recovery

Approximately 0.2 ml of a suspension of bacteria cells (10^{10}/ml) should be introduced into the lateral tail vein of each animal 180 min prior to killing. The mice should be killed by dislocation of the neck. The livers should be removed, suspended in 10 ml ice-cold saline, and homogenized. After initial centrifugation for 10 min at 200g, the supernatant of the homogenate should be centrifuged 2 × 10 min at 7000g, and the centrifuge resuspended in 10 ml of saline solution.

Cell Plating for Survival and Mutant Determinations

The mutant counts should be performed for each animal individually. Four plates of minimal agar with biotin should be spread with 0.25 ml of the undiluted samples from each animal.

The total bacterial counts present in the animals should be determined for each dose or control group as a whole. Dilutions of 10^{-5} and 10^{-6} of samples should be spread on five and four nutrient agar (NA) plates, respectively. No accepted methods of data analysis exist in the literature. However, statistical treatment of mutant frequencies per animal appears appropriate.

References

Ames, B. N., Gurney, E. G., Miller, J. A., and Bartsch, H.: Carcinogens as frameshift mutagens: Metabolites and derivatives of 2-acetylaminofluorene and other aromatic amine carcinogens. *Proc. Natl. Acad. Sci., U.S.A.* **69**:3128–3132, 1972.

Ames, B. N., Lee, F. D., and Durston, W. E.: An improved bacterial test system for the detection and classification of mutagens and carcinogens. *Proc. Natl. Acad. Sci. U.S.A.* **70**:782–786, 1973a.

Ames, B. N., Durston, W. E., Yamasaki, E., and Lee, F. D.: Carcinogens are mutagens: A simple test system combining liver homogenates for activation and bacteria for detection. *Proc. Natl. Acad. Sci., U.S.A.* **70**:2281–2285, 1973b.

Ames, B. N., McCann, J., and Yamasaki, E.: Methods for detecting carcinogens and mutagens with the *Salmonella*/mammalian-microsome mutagenicity test. *Mutat. Res.* **31**:347–364, 1975.

Finney, D. J.: *Probit Analysis*. Cambridge University Press, 1971.

Gabridge, M. G., and Legator, M. S.: A host-mediated microbial assay for the detection of mutagenic compounds. *Proc. Soc. Exp. Biol. Med.* **130**:831–834, 1969.

McCann, J., Springarn, N. E., Kobori, J., and Ames, B. N.: Detection of carcinogens as mutagens: Bacterial tester strains with R factor plasmids. *Proc. Natl. Acad. Sci. U.S.A.* **72**:979–983, 1975a.

McCann, J., Choi, E., Yamasaki, E. and Ames, B. N.: Detection of carcinogens as mutagens in the *Salmonella*/microsome test: Assay of 300 chemicals. *Proc. Natl. Acad. Sci. U.S.A.* **72**:5135–5139, 1975b.

Mohn, G., Ellenberger, J., McGregor, D., and Merker, H. J.: Mutagenicity studies in microorganisms *in vitro* with extracts of mammalian organs, and with the host-mediated assay. *Mutat. Res.* **29**:221–223, 1975.

Vogel, H. J., and Bonner, D. M.: Acetylornithinase of *E. coli* partial purification and some properties. *J. Biol. Chem.* **218**:97–106, 1956.

PROTOCOL 13

BONE MARROW CYTOGENETIC ANALYSIS IN RATS

One means of detecting *in vivo* genetic activity is to examine mitotically active cells that have been arrested at metaphase for structural changes and rearrangement of their chromosomes. The occurrence of such chromosome aberrations correlates well with the administration of known mutagens to animals and thus may serve as an indicator of possible mutagenic potential of test articles.

The protocol that follows has been successfully used by many laboratories and adheres closely to the guidelines established by the *Ad Hoc* Committee on Chromosome Methodologies in Mutagen Testing (1972). Although this protocol is for testing with rats, the CD-1 strain of mice or Chinese hamsters may be used as well.

Materials

Adult male Sprague–Dawley rats of the CRL:COBS CD(SD)BR outbred strain or similar strain should be purchased from a reputable dealer. This healthy, random-bred strain maximizes genetic heterogeneity and at the same time assures access to a common source.

TEM at 1.0 mg/kg may be used as the positive control article administered via a single IP injection. The negative control article should consist of the solvent or vehicle used for the test article administered by the same route as, and concurrently with, the test article.

Experimental Design

Animal Husbandry

Animals should be group housed up to four rats per cage with a commercial diet and water available *ad libitum* unless contraindicated by the particular experimental design.

Animals should be assigned to study groups at random. Prior to study initiation, animals should be weighed to calculate dose levels; the volume of test article administered per animal should be established using the mean animal weight unless there is significant variation among individuals, in which case individual calculations should be made. Animals should be uniquely identified by either ear tag or ear punch, and dose or treatment groups by cage card.

Sanitary cages and bedding should be used. Personnel handling animals or working within the animal facilities should be required to wear suitable protective garments. When appropriate, individuals with respiratory or other overt infections should be excluded from the animal facilities.

Dose Selection (see Appendix B)

If acute toxicity information (e.g., LD_{50}) is available, it may be used to determine dose levels. If it is not available, dose levels can be determined using five groups of six animals each in a toxicity study. The LD_{50} should be determined statistically by probit analysis (Finney, 1971). In the event that an LD_{50} cannot be determined because the test article is nontoxic, the doses for the mutation studies should be selected as high in relation to conditions of human use.

Once the LD_{50} has been approximated, the highest dose for acute studies can be arbitrarily selected as one-eighth to one-tenth the LD_{50}. One-third and one-tenth of the high dose should normally be used as the intermediate and low dose levels, respectively. Subchronic doses should be identical to those used in the acute studies unless contraindicated. An attempt should be made in mutagenesis studies as well as other toxicology work to evaluate the extremes of dosage as well as values close to the use level.

Route of Administration

The route of administration should be by oral gavage. In the event that test article characteristics preclude oral gavage, IP injection should be employed. These routes of administration are the most common routes of administration for this test procedure.

Methodology

The basic design of the test is shown below. Both acute (single dose) and subchronic (five consecutive doses) sequences should be provided. Either sequence may be run individually, but in general, the use of both sequences provides more definitive results. A total of 136 animals should be used in the test—104 for the acute study and 32 for the subchronic study, as outlined.

Treatment	Acute study			Subchronic studies	Total animals
	Number of animals killed after dosing			Five exposures 24 hr apart—Number of animals killed 6 hr after last exposure	
	6 hr	24 hr	48 hr		
High level	8	8	8	8	32
Intermediate level	8	8	8	8	32
Low level	8	8	8	8	32
Positive control[a]	—	8	—	—	8
Negative control	8	8	8	8	32

[a] Positive controls are not necessary in subchronic studies if acute studies are conducted concurrently, since the positive control will be administered on an acute, one-time basis only. TEM (1.0 mg/kg) should be used as the positive control article.

Three hours prior to kill, the animals should be injected IP with 4.0 mg/kg of colchicine. The animals should be killed with CO_2 at the times indicated in the table. The adhering soft tissue and epiphyses of both tibiae should be removed according to the method of Legator et al. (1969). The marrow should be aspirated from the bone and transferred to Hanks' balanced salt solution. The marrow button should be collected by centrifugation and resuspended in 0.075 M KCl. Centrifugation should be repeated, and the pellet resuspended in fixative (methanol:acetic acid, 3:1). The fixative should be changed after one-half hour and the cells left overnight at 4°C.

Cells in fixative should be dropped onto glass slides and air-dried. Spreads should be stained with 10% Giemsa at pH 6.8.

Slides should be coded and scored for chromosomal aberrations. The complete list of aberrations to be scored for is as follows:

chromatid gap	pulverized chromosome
chromatid break	pulverized chromosomes
chromosome gap	pulverized cells
chromosome break	complex rearrangement
chromatid deletion	ring chromosome
fragment	dicentric chromosome
acentric fragment	minute chromosome
translocation	greater than 10 aberrations
triradial	polyploid
quadriradial	aneuploid

Routinely, 50 spreads should be read for each animal. Depending on the frequency of scorable metaphase cells, it may be necessary to prepare additional slides from the original cell suspension. The location of cells bearing aberrations should be identified by the use of coordinates on the mechanical stage. A mitotic index based on at least 500 cells counted should also be recorded. It should be calculated by scoring the number of cells in mitosis per 500 cells on each slide read.

For control of bias, all slides should be coded prior to scoring and scored blind.

Evaluation Criteria

General guidelines have been established to serve as aids in determining the meaning of bone marrow chromosomal aberrations. Basically, an attempt should be made to establish whether a test article or its metabolites can interact with chromosomes to produce gross lesions or changes in chromosome numbers, and whether these are of a type which can survive more than one mitotic cycle of the cell. All aberration figures detected by this assay result from breaks in the chromatin which either fail to repair or repair in atypical combinations. The cell transit time for bone marrow is normally 8–24 hr. The assay design is such that bone marrow samples should be taken at 6, 24, and 48 hr after an acute administration of the test article to permit detection of chromosome aberrations in cells that have been delayed in their progression through the mitotic cycle.

One would anticipate that many of the cells bearing breaks or reunion figures would be eliminated after their first mitotic division, and as a corollary, that those cells which survive the first anaphase would primarily bear balanced lesions. The detection of these lesions and hence a complete risk evaluation must usually rely on additional testing. In general, a cell bearing configurations such as small deletions or reciprocal translocations may be perpetuated and therefore constitute a greater risk to the individual than large deletions or complex rearrangements.

The type of aberration, its frequency, and its correlation to dose in a given time period should all be considered in evaluating a test article as being mutagenically positive or negative. Gaps should not be counted as significant aberrations unless they are present in a much higher than usual frequency. Open breaks should be considered as indicators of genetic damage, as should configurations resulting from the repair of breaks. The latter includes translocations, multiradials, rings, multicentrics, etc. Reunion figures such as these should be weighted slightly higher than breaks, since they usually result from more than one break and may lead to stable configurations.

The number of aberrations per cell should also be considered significant; cells with more than one aberration should be considered indicative of more genetic damage than those containing evidence of single events. Consistent variations from the euploid number should also be considered in the evaluation of matagenic potential.

Frequently, one is unable to locate 50 suitable metaphase spreads for each animal even after preparing additional slides. Possible causes for this appear to be related to cytoxic effects which alter the duration of the cell cycle, kill the cell, or cause clumping of the chromosomes. Additional information can be gained from the mitotic index, which also appears to reflect cytotoxic effects.

Comparison with a concurrent negative control that happens to show an unusually low frequency of aberrations can suggest undue statistical significance; Therefore, treatment data should also be considered against historical control data.

References

Chromosome methodologies in mutagen testing: Report of the *Ad Hoc* Committee of the Environmental Mutagen Society and the Institute for Medical Research. *Toxicol. Appl. Pharmacol.* **22**:269–275, 1972.

Dean, B. J.: Chemical-induced chromosome damage. *Lab Animal* **3**:157–174, 1969.
Legator, M. S., Palmer, K. A., Green, S., and Peterson, K. W.: Cytogenetic studies in rats of cyclohexalamine, a metabolite of cyclamate. *Science* **165**:1139–1140, 1969.

PROTOCOL 14

DOMINANT LETHAL ASSAY IN RATS

The dominant lethal assay is an *in vivo* test to determine the germ cell risk from a suspected mutagen once its genetic activity has been clearly demonstrated using *in vitro* or other suitable mutagenicity screens. The following protocol describes the dominant lethal assay for germ cell risk assessment using rats as the test system. Assays should be performed at three dose levels established from toxicologic LD_{50} determinations. Positive and negative controls also should be conducted.

The standard procedure consists of administering a test article to male rats for 5 days, followed by sequential mating of the dosed males to new sets of females each week for a sufficient number of weeks to cover the total spermatogenic cycle.

This assay may also be performed using mice or hamsters.

Materials

Fifty adult male and 700 virgin female rats of the CRL:COBS CD(SD)BR outbred strain or similar strain should be purchased from a reputable dealer. This healthy, randomly bred strain has been selected to maximize genetic heterogeneity and at the same time assure access to a common source.

TEM at 0.3 mg/kg may be used as the positive control article, administered via a single IP injection. The negative control article should consist of the solvent or vehicle used for the test article, administered by the same route as, and concurrently with, the test article.

Experimental Design

Animal Husbandry

Animals should be group housed up to four rats per cage with a commercial diet and water available *ad libitum* unless contraindicated by the particular experimental design.

For control of bias, animals should be assigned to study groups at random. Prior to study initiation, animals should be weighted and an overall mean weight determined. The volume of test article to be administered per animal should be established using this mean weight unless there is significant variation among individuals, in which case individual calculations should be made. Animals should be uniquely identified by either ear tag or ear punch, and treatment groups identified by cage card.

Sanitary cages and bedding should be used. Personnel handling animals or working within the animal facilities should be required to wear suitable protective garments. When appropriate, individuals with respiratory and other overt infections should be excluded from the animal facilities.

Dose Selection (see Appendix B)

If acute toxicity information (e.g., LD_{50}) is available, it can be used to determine dose levels. If it is not available, dose levels can be determined using five groups of six animals each in a toxicity study. The LD_{50} can be determined statistically by probit analysis (Finney, 1971). In the event that an LD_{50} cannot be determined because the test article is nontoxic, the highest dose is generally 5 g or 5 ml/kg.

Normally, if the LD_{50} can be approximated, the highest dose for acute studies should be arbitrarily selected as one-tenth the LD_{50}. One-third and one-tenth this high dose should be used as the intermediate and low dose levels, respectively.

Route of Administration

Ordinarily, oral gavage should be chosen as the route of administration. In the event that test article characteristics preclude oral gavage, IP injection should be employed. These routes of administration are the most common for this test procedure.

Methodology

The test and negative control articles should be administered to each of 10 males per dose level daily for 5 days. On day 5 of the dosing schedule, the animals serving as the positive control should receive an acute IP injection of TEM.

Following treatment the males are sequentially mated to two females per week for 7 weeks. This number of females provides an adequate number of implantations per group per week. After the mating period, the females should be removed and housed in other cages until killed. The males should be rested on Saturday and Sunday, and two new females introduced into each cage on Monday. Conception generally takes place in more than 90% of the females by Friday, and a 2-day rest is beneficial to the males with regard to subsequent weekly matings.

Females should be killed using CO_2 at 14 days after the midweek of mating. At necropsy the uteri should be examined, and the number of corpora lutea and living and dead implantations counted for each pregnant female. From this data base the following parameters should be calculated and evaluated by appropriate statistical methods:

1. *Fertility index.* The fertility index should be computed as number of pregnant females per number of mated females.
2. *Total number of implantations.*
3. *Total number of corpora lutea.*
4. *Preimplantation losses.* Preimplantation losses should be computed for each female by subtracting the number of implantations from the number of corpora lutea. This is not usually performed in mice.
5. *Dead implantations.*
6. *Proportion of females with one or more dead implantations.* The proportion of females with one or more dead implantations should be computed as number of females with dead implants per number of pregnant females.
7. *Proportion of females with two or more dead implantations.* This proportion of females with two or more dead implantations should be computed as number of females with two or more dead implants per number of pregnant females.
8. *Dead implants/total implants.* Dead implants/total implants should be computed for each female.

It should be noted that each parameter and each week should be evaluated independently.

References

Epstein, S., Arnold, E., Andrea, J., Bass, W., and Bishop, Y.: Detection of chemical mutagens by the dominant lethal assay in the mouse. *Toxicol. Appl. Pharmacol.* **23**:288–325, 1972.
Finney, D. J.: *Probit Analysis.* Cambridge University Press, 1971.

PROTOCOL 15

HERITABLE TRANSLOCATION ASSAY IN MICE

The objective of this assay is to determine the ability of an article to induce reciprocal translocations between chromosomes in germ cells of treated male mice. In the HTA, the presence of induced translocations can be detected by mating the male F_1 progeny of treated males with unrelated, untreated females and scoring for a reduction in the number of viable fetuses. This reduction (semisterility) may be directly attributable to the F_1 male being a translocation heterozygote, that is, the male having inherited a pair of nonhomologous chromosomes previously involved in a reciprocal translocation, thus producing duplication/deletion gametes at the subsequent meiosis. Such gametes may be nonviable or may result in early embryonic death. The presence of reciprocal translocations can be verified cytogenetically by the presence of translocation figures in meiotic metaphase. The reciprocal translocation assay provides a sensitive method for estimating heritable genetic damage resulting from chromosome breakage.

The primary advantage of the HTA over other *in vivo* tests, such as the dominant lethal assay or bone marrow cytogenetic assay, is that it measures a transmissible lesion. Genetic alterations which can pass through spermatogenic stages and therefore be transmitted to subsequent generations are more damaging to the gene pool than dominant lethal and other nontransmissible chromosome alterations.

Materials

CD-1 randomly bred Swiss albino mice are highly suitable animals for this assay, having been used successfully by several investigators to test known mutagens. In fact, the predictions by Snell (1933 and 1946) that semisterility resulted from reciprocal translocation were postulated and cytogenetically confirmed from experiments using this mouse as the experimental animal.

A positive control can be run concurrently with the test article. Otherwise, a historical control based on the results of other recent studies should be included.

Negative controls should always be run concurrently with the study; they should consist of the solvent or vehicle used for the test article. When applicable, TEM (0.3 mg/kg) may be used as the positive control article, administered via a single IP injection. At this dose level, TEM exerts a pronounced effect on postmeiotic cells.

Experimental Design

Animal Husbandry

Animals should be group housed up to 15 mice per cage with a commercial diet and water available *ad libitum* unless contraindicated by the particular experimental design.

Animals should be assigned to study groups at random. Prior to study initiation, animals should be weighed, and the volume of test article to be administered per animal established using a mean weight unless there is significant variation among individuals, in which case individual calculations should be made. Animals should be uniquely identified by either ear tag or ear punch; treatment groups should be identified by cage card.

Sanitary cages and bedding should be used. Personnel handling animals or working within the animal facilities should be required to wear suitable protective garments. When appropriate, individuals with respiratory or other overt infections should be excluded from the animal facilities.

Dose Selection (see Appendix B)

If acute toxicity information (e.g., LD_{50} is available, it should be used to determine dose levels as described below. If acute toxicity information is not available, dose levels should be determined using five groups of six animals each in a toxicity study. The LD_{50} should be determined statistically by probit analysis (Finney, 1971). In the event that an LD_{50} cannot be determined because the test article is nontoxic, the highest dose is generally 5 g or 5 ml/kg.

Once the LD_{50} has been approximated, the high dose in this study should be arbitrarily selected as one-eighth the LD_{50}, with the medium dose being one-third the high dose, and the low dose being one-tenth the high dose.

The route of administration should be oral gavage unless contraindicated. This is the most common route of administration used for this test procedure.

Methodology

The HTA requires approximately 50 weeks as outlined in Figure 8.6. Thirty animals per dose level should receive predetermined doses of the test and control articles daily for 7 weeks. Approximately 7 weeks are required for the progeny of a stem cell to traverse the entire period of spermatogenesis and emerge as mature sperm. Since, theoretically, genetic damage can be induced in premeiotic as well as postmeiotic stages, all parts of the gametogenic pathway should be examined for possible reciprocal translocations, or at least to ensure that cells in each stage have been exposed to the test article. This latter goal can be achieved with surety if the males are dosed for 7 weeks prior to mating.

After this dosing period, each male should be mated with two females. From the offspring of each group, 200 healthy males should be selected. When they reach reproductive maturity, each of these F_1 males should be mated to three virgin females for 1 week.

Approximately 2 weeks after mating, the females should be killed with

CO_2, surgically opened, and scored for the number of living fetuses and resorbed embryos present in their uteri. Semisterility should be determined from this information on the basis of a specific set of criteria.

When a male is able to produce 10 or more living embryos with one of the three females, the male is considered to be fertile and no further matings are necessary. A male which does not meet this criterion should be remated with three females. This sequential mating of the F_1 males should be repeated up to three times. The males which are not able to produce 10 or more living embryos in one of their matings should be considered sterile or semisterile. They should be killed and their gonadal cells analyzed for evidence of translocations. The basic method should be that of Ford and Evans (1969), in which the testes are teased apart with fine forceps and then subjected to a mild trypsinization. The released cells are washed, processed, and air-dried onto slides. Stained slides are examined for the presence of translocation figures indicative of translocation heterozygosity among them (100 cells per animal).

References

Finney, D. J.: *Probit Analysis*. Cambridge University Press, 1971.

Ford, C. E., and Evans, E. P.: Meiotic preparation from mammalian testes. In *Comparative Mammalian Cytogenetics* (K. Benirschke, ed.), Springer-Verlag, New York, pp. 461–464, 1969.

Snell, G. D.: X-ray sterility in the male house mouse. *J. Exp. Zool.* **65:**421–441, 1933.

Snell, G. D.: Analysis of translocation in the mouse. *Genetics* **31:**157–180, 1946.

PROTOCOL 16

MOUSE MICRONUCLEUS ASSAY

The objective of this study is to evaluate a test article for clastogenic activity in polychromatic erythrocyte (PCE) stem cells in treated mice. The micronucleus test can serve as a rapid screen for clastogenic agents and test articles which interfere with normal mitotic cell division (Heddle, 1973; Jenssen *et al.*, 1974; Schmid, 1975; Maier and Schmid, 1976). Micronuclei are believed to be formed from chromosomes or chromosome fragments left behind during anaphase and can be scored during interphase because they persist (Schmid, 1975). Thus, the time involved in searching for metaphase spreads in treated cell populations is eliminated. Test articles affecting spindle fiber function or formation, as well as clastogenic agents, can be detected through micronucleus induction (Schmid, 1975).

Materials

Adult male and female mice, strain CD-1, from a randomly bred closed colony should be purchased from a reputable dealer. This healthy, random-bred strain has been selected to maximize genetic heterogeneity and at the same time assure access to a common source.

TEM at 1.0 mg/kg may be used as the positive control article and should be administered via a split dose IP injection. The negative control article should consist of the solvent or vehicle used for the test article and should be administered by the same route as, and concurrently with, the test article in volumes equal to the maximum amount administered to the experimental animals.

Experimental Design

Animal Husbandry

Animals should be group housed up to 15 mice per cage. A commercial diet and water should be available *ad libitum* unless contraindicated by the particular experimental design.

Thirty-two outbred mice, four males and four females per dose level, should be assigned to study groups at random. Prior to study initiation, animals should be weighed to calculate dose levels. The volume of test article administered per animal should be established using a mean weight unless there is significant variation among individuals, in which case individual calculations should be made. Animals should be uniquely identified by ear tag or ear punch, and dose or treatment groups identified by cage card.

Sanitary cages and bedding should be used. Personnel handling animals or working within the animal facilities should be required to wear suitable protective garments. When appropriate, individuals with respiratory or other overt infections should be excluded from the animal facilities.

Dose Selection (see Appendix B)

If acute toxicity information (e.g., LD_{50}) is available, it can be used to determine dose levels. If it is not available, dose levels can be determined using five groups of six animals each in a toxicity study. The LD_{50} should be determined statistically by probit analysis (Finney, 1971). In the event that an LD_{50} cannot be determined because the test article is nontoxic, the doses for the mutation studies should be selected as high in relation to conditions of human use.

Once the LD_{50} has been approximated, the high dose generally should be selected as one-half the LD_{50}, with the low dose being one-tenth the high dose. An attempt should be made in mutagenesis studies, as well as other toxicology work, to evaluate the extremes of dosage as well as values close to the use level.

Dosing Schedule and Route of Administration

Previously, a subchronic dosing regimen was used consisting of two administrations approximately 24 hr apart. Treatment convered a 30-hr period, or two cell cycles (Schmid, 1975). More recently, the dosing scheme has been modified to administer the test article in an acute dose followed by micronuclei sampling at 6, 24, and 48 hr after the exposure.

Oral gavage should be employed as the route of administration. In the event that test article characteristics preclude oral gavage, IP injection should be employed. These routes of administration are the most common routes of administration for this test procedure.

Extraction of Bone Marrow

Six hours after the last dose, animals should be killed with CO_2, and the adhering soft tissue and epiphyses of both tibiae removed. The marrow should be aspirated from the bone and transferred to centrifuge tubes containing 5 ml FCS (one tube for each animal).

Preparation of Slides

Following centrifugation to pellet the tissue, the supernatant should be drawn off, cells resuspended in a drop of serum, and suspension spread on slides and air-dried. The slides should then be fixed in methanol, stained in May–Gruenwald solution followed by Giemsa, and rinsed in distilled water.

Screening the Slides

A thousand PCEs per animal should be scored. The frequency of micronucleated cells should be expressed as percent micronucleated cells based on the total PCEs present in the scored optic field. The normal frequency of micronuclei in the CD-1 mouse strain is approximately 0.2%. The frequencies of other bone marrow cell types should be recorded for analysis of cytotoxic effects (reduced production of specific cell types).

Evaluation Criteria

In tests performed for this evaluation, only PCEs should be scored for micronuclei. Mature erythrocytes and other cells in the field should be recorded but not scored for micronuclei.

The dose levels should be established to ensure that a nontoxic level of the test article is scored. Dose response data are not necessary to define a test article as active. Responses considered active should be assumed to reflect clastogenic and related activities of test articles. Agents which break chromosomes and induce nondisjunction and other events which produce structural or numerical changes in chromosomes can produce micronuclei.

The data generated in this study should be analyzed by a two-tailed t-test (Finney, 1971). Individual animal results should be used as data points in the analysis. The set of micronuclei frequencies among the controls should be compared with the set for each treatment level. Male and female animal data should be combined unless there appears to be a sex difference, in which case the data should be analyzed separately. Increases above the negative control frequency that are significant at $p < 0.01$ should be considered indicative of an active agent.

For control of bias, all slides should be coded prior to scoring, and scored blind.

References

Finney, D. J.: *Probit Analysis*. Cambridge University Press, 1971.

Heddle, J.: A rapid *in vitro* test for chromosomal damage. *Mutat. Res.* **18**:187–190, 1973.

Jenssen, D., Ramel, C., and Gothe, R.: The induction of micronuclei by frameshift mutagens at the time of nucleus expulsion in mouse erythroblasts. *Mutat. Res.* **26**:553–555, 1974.

Maier, R., and Schmid, W.: Ten model mutagens evaluated by the micronucleus test. *Mutat. Res.* **40**:325–338, 1976.

Schmid, W.: The micronucleus test. *Mutat. Res.* **31**:9–15, 1975.

PROTOCOL 17

MOUSE SPERMHEAD ABNORMALITIES

The objective of this assay is to determine whether a test article can disrupt the normal morphology of the mammalian spermhead. Abnormalities observed in the spermhead presumably occur during spermatogenesis, as once the spermhead develops its shape it is extremely stable.

Materials

Adult hybrid male mice (B6C3F1/CrlBR) or a similar strain should be purchased from a reputable dealer. Such hybrids are suitable for use in this assay, since reasonably low and consistent background levels of abnormal sperm have been observed.

EMS at 200 mg/kg may be used as the positive control article, administered acutely via IP injection. The negative control article should consist of the solvent or vehicle used for the test article and should be administered by the same route as, and concurrently with, the test article.

Experimental Design

Animal Husbandry

Animals can be group housed up to 15 mice per cage. A commercial diet and water should be available *ad libitum* unless contraindicated by the particular experimental design.

Animals should be assigned at random to study groups. Prior to study initiation, animals should be weighed, and the volume of test article to be administered per animal established using the mean weight unless there is significant variation among individuals, in which case individual calculations should be made. Animals should be uniquely identified by either ear tag or ear punch, and treatment groups identified by cage card.

Sanitary cages and bedding should be used. Personnel handling animals or working within the animal facilities should be required to wear suitable protective garments. When appropriate, individuals with respiratory or other overt infections should be excluded from the animal facilities.

Dose Selection (see Appendix B)

If acute toxicity information (e.g., LD_{50}) is available, it can be used to determine dose levels. If acute toxicity information is not available, dose levels can be determined using five groups of six animals each in a toxicity study. The LD_{50} should be determined statisitically by probit analysis (Finney, 1971). In the event that an LD_{50} cannot be determined because the test article is nontoxic, the doses should be selected as high in relation to conditions of human exposure.

Once the LD_{50} has been approximated, the high dose generally should be selected as one-eighth the LD_{50}, with the low dose being one-eighth the high dose. Concentrations which inhibit spermatogenesis should be avoided.

Dosing Schedule and Route of Administration

Normally a subchronic dose regimen should be used consisting of five daily administrations. Oral gavage should be employed as the route of administration. In the event that test article characteristics preclude oral gavage. IP injection should be employed. These routes of exposure are the most common routes of administration for this test procedure.

The animals should be killed at 1, 3, and 5 weeks following the end of exposure, and their sperm examined. The sperm thus sampled presumably will be derived from cells exposed during spermatogenesis as sperm, spermatids, and spermatocytes. At each dose level, 30 animals should be used (10 for each time period).

Preparation of Sperm Suspension

The animals should be killed with CO_2. The caudae epididymides should be dissected and placed in a centrifuge tube containing 3 ml of 0.9% saline solution. The contents of the tube should then be transferred to a petri dish, and the caudae cut into several small pieces with small scissors. The resulting suspension should be gently pipetted five to six times up and down in a 5- or 10-ml pipette. The sperm solution should be filtered through an 80-μm silk mesh to remove tissue fragments, and 0.5 ml of the filtrate transferred to a centrifuge tube to which 0.05 ml of 1% Eosin Y is added. The solution should be gently agitated. Slides should be made by placing one drop of the stained solution on a slide and spreading by three passes of another slide. The slides should then be air-dried and mounted with Permount. At least 1000 sperm should be examined per animal.

For control of bias, all slides should be coded prior to scoring and scored blind. Figure 9.3 illustrates normal and abnormal spermheads from both mice and rats. The procedure in rats is essentially identical to the mouse study design.

Evaluation Criteria

The negative control should consist of the solvent or vehicle of the test article administered to the animals on the same schedule as the test article. This control gives a reference point to which the test data can be compared.

If the negative control is within the normal range, a test article that produces more abnormalities than fall within the 90% probability range of the historical controls, as determined by probit analysis (Finney, 1971), should be considered positive in this test. Dose-response data are not necessary to define a test article as active.

Since the folded-head type of abnormality might not represent genetic damage, this aberration should not be included in the evaluation of the test article.

References

Finney, D. J.: *Probit Analysis*. Cambridge University Press, 1971.
Wyrobek, A. J., and Bruce, W. R.: Chemical induction of sperm abnormalities in mice. Proc. Natl. Acad. Sci. **U.S.A. 72:**4425–4429, 1975.

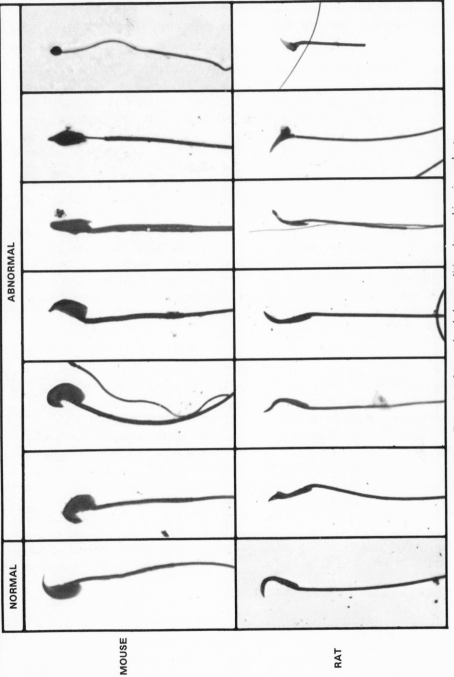

FIGURE 9.3. Examples of sperm-head abnormalities observed in mice and rats.

SOMATIC MUTATION ASSAY IN THE MOUSE (SPOT TEST)

The objective of this study is to detect somatic gene mutations *in vivo* using specially derived strains of mice. (Fahrig, 1975; Russell and Major, 1975; Russell, 1977).

Mutations are initially induced in melanocyte precursor cells in developing embryos that are heterozygous at five specific coat color loci (see Figure 8.4). Melanocyte precursor cells carrying a mutation at any wild-type allele of the five coat color loci will develop into clones of "mutant" melanocytes. Such clones can be readily recognized as coat color mosaic patches on heterozygous mice. Since each developing embryo contains approximately 150 to 200 melanocyte precursor cells, a relatively small number of animals is required to provide reliable data in this *in vivo* somatic mutation assay.

The method basically consists of treating females carrying 9- to 11-day embryos that are heterozygous at specific coat-color loci and subsequently examining the young, after birth, for any mosaic patches (i.e., clones of mutant cells) in differentiated fur. These spots could result from a number of mechanisms, i.e., point mutations, chromosomal deletions, whole chromosome loss (caused by nondisjunction and other mechanisms), and mitotic crossing-over.

Materials

Animals

The T strain of mice used for this assay may be obtained from Dr. L. Russell of the Oak Ridge National Laboratory, Oak Ridge, TN; C57Bl/6J female mice, or a similar strain, are also used for this assay, and should be purchased from a reputable dealer. These strains are used because of genotypic requirements of this assay (Figure 5.2).

Control Articles

Benzo[a]pyrene (150 mg/kg) or ethyl nitrosourea (50 mg/kg) can be used as the positive control article, administered via a single IP injection on day 10 of pregnancy. The negative control article should consist of the solvent or vehicle used for the test article, administered by the same route as, and concurrently with, the test article.

Experimental Design

Animal Husbandry

Animals should be group housed up to 15 mice per cage. A commercial diet of approximately 6 and 9% fat should be fed to the T strain and the

C57Bl/6J female mice, respectively. Food and water should be available *ad libitum* unless contraindicated by the particular experimental design.

Animals should be assigned to study groups at random. Prior to test article administration, animals should be weighed, and the volume of test article to be administered per animal established using the mean weight unless there is significant variation among individuals, in which case individual calculations should be made. Animals should be uniquely identified by either ear tag or ear punch, and treatment groups identified by cage card.

Sanitary cages and bedding should be used. Personnel handling animals or working within the animal facilities should be required to wear suitable protective garments. When appropriate, individuals with respiratory or other overt infections should be excluded from the animal facilities.

Dose Selection

If acute toxicity information (e.g., LD_{50}) is available, it should be used to determine dose levels as described below. If it is not available, dose levels should be determined using five groups of six animals each in a toxicity study. The LD_{50} should be determined statistically by probit analysis (Finney, 1971). In the event that an LD_{50} cannot be determined because the test article is nontoxic, the doses should be selected as high in relation to conditions of human exposure.

Normally, when an LD_{50} can be determined, the highest dose should be arbitrarily selected as one-tenth the LD_{50}. One-third and one-tenth this high dose should normally be used as the intermediate and low dose levels, respectively. An attempt should be made in mutagenesis studies as well as other toxicology work to evaluate the extremes of dosage as well as values close to the use level.

Route of Administration

The route of administration employed should be oral gavage. In the event that test article characteristics preclude oral gavage, IP injection should be employed. These routes of administration have been selected because they are the most common routes of administration for this test procedure.

Treatment

Animals should be mated according to the following scheme: T-strain males × C57Bl/6J females. During the mating period, the females should be checked for copulation plugs. Following mating, females should be removed from the mating cages, ear tagged, and randomly assigned to one of the control or treatment groups. Fifty plugged females should be included in each control (negative and positive) and test group. The females should be treated on days 9 to 11 of pregnancy. The number of pigments in developing embryos should be expected to be 150 to 200 at that time.

Spot Observation

All exposed embryos should be allowed to develop to birth. At birth, the litter size should be determined, and external morphologies examined for any gross abnormalities. The newborns should be scored for spots on day 14 and at the time of weaning. White spots near the ventral midline should be recorded separately from all other coat color spots. White mid-ventral spots represent melanocyte toxicity and not mutation induction. Animals with mutant spots should be retained frozen for 5 years.

Frequencies of coat color spots in treated and control groups should be compared according to Fisher's exact test for statistical significance of mutation frequencies. Increases over background that are statistically significant at the 1% level should also be considered biologically significant.

References

Fahrig, R.: A mammalian spot test: Induction of genetic alterations in pigment cells of mouse embryos with X-rays and chemical mutations. *Mol. Gen. Genet.* **138**:309–314, 1975.

Finney, D. J.: *Probit Analysis.* Cambridge University Press, 1971.

Russell, L. B.: Validation of the *in vivo* somatic mutation method in the mouse as a prescreen for germinal point mutations. *Arch. Toxicol.* **38**:75–85, 1977.

Russell, L. B., and Major, M. H.: Radiation-induced presumed somatic mutations in the house mouse. *Genetics* **42**:161–175, 1975.

PROTOCOL 19

TEST FOR LOSS OF X AND Y CHROMOSOMES IN *D. MELANOGASTER*

The test for loss of X and Y chromosomes in *D. melanogaster* is a rapid, one-generation screen for test articles that cause breakage in chromosomes. The test is designed to monitor loss of genetic markers on the X and Y chromosomes that is interpreted as being due to breaks that are not rejoined. This test is useful as a supplement to the sex-linked recessive lethal test.

Materials

The stocks utilized for this test should be chosen because of unique chromosome structure and visible markers. They should be maintained as permanent cultures and visually examined on a routine basis for any indication of genetic breakdown.

Tetraethyleneimino-1,4-benzoquinone in 1% sucrose solution may be administered as the positive control as a single, 24-hr exposure on the final day of the dosing schedule. The males dosed should be of the same age and stock as those used for the test article. The solvent or vehicle used for the test article should be used as the negative control and be administered concurrently with the test article.

Drosophila cultures should be maintained at ~25°C in disposable vials. The culture medium to be used can be Carolina Biological Instant *Drosophila* medium (Formula 4-24, without dyes or the equivalent). Flies should be immobilized with filtered CO_2 for handling.

Experimental Design

Solubility Testing

Each test article should be tested for solubility; the primary choice for a solvent should be distilled water. If a test article is insoluble in water, it should be dissolved in DMSO, ethanol, acetic acid, or acetone and then diluted. The final concentration of solvent (other than water) should be $\leq 2\%$ of the feeding solution.

Palatability and Toxicity Testing

A test for palatability and toxicity should be performed by preparing a test article in distilled water containing sucrose and any necessary solvents. The solution should be administered by the following method mofified from Lewis and Bacher (1968). A piece of chemically inert glass filter paper lining a shell vial should be saturated with 1.5 ml of the test solution. Fifty adult males should be placed in each feeding vial and observed for feeding behavior and toxic effects. To ensure uptake it may be necessary to use concentrations of sucrose varying between 1 and 5% and to vary the feeding time between 1 and 3 days. It will be useful at this point to include food coloring in the feeding solutions of certain vials, as the coloring is readily distinguishable in the gut and feces of *Drosophila* and it can be easily determined that the flies have ingested the test article. Vials containing the coloring agent should not be used for assessment of toxicity. Should it be necessary to extend the dosing period beyond 24 hr, the test article should be reprepared and the flies transferred to a new feeding vial.

Fertility Testing

Once the solubility and toxicity have been established, a small pilot study should be conduced to determine the fertility of the treated males. Selected concentrations of the test article should be prepared, and the dosed males put through the same mating program as that to be used in the actual test to ensure that sufficient numbers of progeny will be available for analysis.

Test Size

The tests should be conducted at three dose levels in conjunction with negative and positive controls. At least 100 treated males should be used for each dose level of the test article and negative and positive controls.

Procedure

Collection of Males

To provide males of approximately uniform age for treatment, males from culture bottles in optimum condition which are produced from timed (1-day) egg samples should be collected as virgins and held for a 2-day maturation period. All males should then be between 2 and 3 days of age and have their full complement of sperm for treatment. They should be randomized by the following unsystemized method. A sufficient number of healthy males for the test should be transferred to a single empty bottle, mixed, parcelled into groups of 50, and held in empty vials for several hours before exposure. This period of removal from a food supply ensures immediate uptake of the feeding solution.

Dosing

The males should be dosed according to the methods described under Palatability and Toxicity Testing, using the concentration and duration of exposure determined in the preliminary studies. The high dose should normally be set at one-half the LD_{50} and the low dose at one-quarter the LD_{50} unless contraindicated by sterility of the dosed males. The maximum concentration administered, if the test article is nontoxic, should generally be 5%.

Mating and Scoring

The mating scheme to be used is given in Figure 9.4. Two- to three-day-old $yB/y+Y$ males should be treated and mass-mated to virgin "Oster, *Inscy*" females, 25 pairs per bottle, for 24 hr. The males should then be

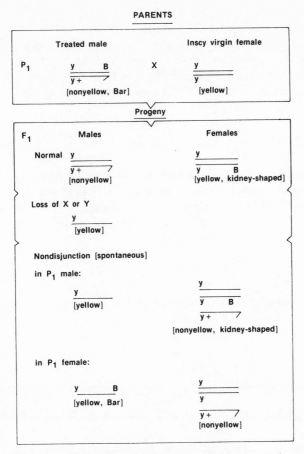

FIGURE 9.4. Mating scheme for the *Drosophila* X or Y chromosome loss assay. Genetic markers: *y*, recessive, yellow-bodied; *y*+, dominant wild-type allele of *y*, non-yellow-bodied; *B*, dominant, narrow-shaped eye when hemizygous and homozygous; kidney-shaped eye when heterozygous. Phenotypes are indicated in parentheses.

removed and the females allowed to lay eggs for 2 days. The F_1 progeny should be scored for the presence of yellow males. As shown in the mating scheme, F_1 yellow males can originate from (1) loss of X or Y in P_1 males, (2) nondisjunction in P_1 males, and (3) nondisjunction in "Oster, *Inscy*" females. However, class (2) can be accounted for by counting its complementary nonyellow females with kidney-shaped eyes, which occur at equal frequency with class (2) males. Class (3) yellow males can be identified by the presence of narrow-shaped Bar eyes. An average of 5000 F_1 progeny should be scored for each dose level.

Data Analysis

The accurate number of X and Y chromosome losses is obtained by subtracting the number of nonyellow females with kidney-shaped eyes and yellow males with narrow-shaped Bar eyes from the total number of yellow males. The frequency of X and Y chromosome loss is calculated as

$$\frac{\text{number of X and Y losses}}{\text{total F}_1 \text{ progeny}} \times 100$$

This frequency is calculated for each dose level and control. The data should be subjected to the Kastenbaum–Bowman statistical test to determine the significance of the results. Generally, an increase of twice the spontaneous frequency in conjunction with a dose-response relationship is considered a positive response.

Reference

Lewis, E. B., and Bacher, F.: Method of feeding ethlmethane sulfonate (EMS) to *Drosophila* males. *Dros. Inf. Service* **48**:193, 1968.

PROTOCOL 20

HERITABLE TRANSLOCATION TEST IN *D. MELANOGASTER*

The test for translocations in *D. melanogaster* monitors reciprocal exchanges between broken fragments of the second and third chromosomes which can pass through meiotic stages and be transmitted to subsequent generations. This test is useful for detecting chromosome breakage ability and is recommended for test articles that have shown positive results in other *Drosophila* assays such as the dominant lethal, X or Y chromosome loss, and sex-linked recessive lethal tests.

Materials

The stocks utilized for this test should be chosen for unique chromosome structure and visible markers. They should be maintained as permanent cultures and visually examined on a routine basis for any indication of genetic breakdown.

TEM in 1% sucrose solution may be administered as the positive control as a single, 24-hr exposure on the final day of the dosing schedule. The males dosed should be of the same age and stock as those used for the test article. The solvent or vehicle used for the test article should be used as the negative control and be administered concurrently with the test article.

Drosophila cultures should be maintained at ~25°C in disposable vials. The culture medium may be Carolina Biological Instant *Drosophila* medium (Formula 4-24, without dyes or the equivalent). Flies should be immobilized with filtered CO_2 for handling.

Experimental Design

Solubility Testing

Each test article should be tested for solubility; the primary choice for a solvent should be distilled water. If a test article is insoluble in water, it should be dissolved in DMSO, ethanol, acetic acid, or acetone and then diluted. The final concentration of solvent (other than water) should be 2% or less.

Palatability and Toxicity Testing

A test for palatability and toxicity should be performed by preparing a test article in distilled water containing sucrose and any necessary solvents. The solution should then be administered by a method modified from Lewis and Bacher (1968). A piece of chemically inert glass filter paper lining a shell vial should be saturated with 1.5 ml of the test solution. Fifty adult males should be placed in each feeding vial and observed for feeding behavior and toxic effects. To ensure uptake it may be necessary to use concentrations of sucrose varying between 1 and 5% and to vary the feeding time between 1 and 3 days. It will be useful at this point to include food coloring in the feeding solutions of certain vials, as the coloring is readily distinguishable in the gut and feces of *Drosophila* and it can easily be determined that the flies have ingested the test article. Vials containing the coloring agent should not be used for assessment of toxicity. Should it be necessary to extend the dosing period beyond 24 hr, the test article should be reprepared and the flies transferred to a new feeding vial.

Fertility Testing

Once the solubility and toxicity have been established, a small pilot study should be conducted to determine the fertility of the treated males. Selected concentrations of the test article should be prepared, and the dosed males put through the same mating sequence as that to be used in the actual test to ensure that sufficient numbers of progeny will be available for analysis.

Test Size

The tests should be conducted at two dose levels in conjunction with negative and positive controls. At least 150 treated males should be used for each dose level of the test article and negative control. A suggested number of chromosomes to be tested and brooding pattern are shown in the following table:

	Chromosomes Tested			
	Brood I	Brood II	Brood III	
Tests	1, 2	3, 4, 5	6, 7, 8	Total
Low concentration	1350	1350	1350	4050
High concentration	1350	1350	1350	4050
Negative control	1350	1350	1350	4050
Positive control	100	100	100	300

Procedure

Collection of Males

To provide males of approximately uniform age for treatment, males from culture bottles in optimum condition which are produced from timed (1-day) egg samples should be collected as virgins and held for a 2-day maturation period. All males should then be between 2 and 3 days of age and should have their full complement of sperm for treatment. Males should be randomized by the following unsystemized method. A sufficient number of healthy males for the test will be transferred to a single empty bottle, mixed, parcelled into groups of 50, and held in empty vials for several hours before exposure. This period of removal from a food supply should ensure immediate uptake of the feeding solution.

Dosing

The males should be dosed according to the methods described under Palatability and Toxicity Testing, using the concentration and duration of exposure determined in the preliminary studies. The high dose should normally be set at one-half the LD_{50} and the low dose at one-quarter the LD_{50} unless contraindicated by sterility of the dosed males. The maximum concentration administered, if the test article is nontoxic, should generally be 5%.

Mating and Scoring

The mating scheme which should be used is given in Figure 9.5. The treated wild-type males should be mated individually to three virgin "Oster, *bw st p^p^"* females. The brooding scheme followed should consist of a 2-3-3-day sequence which samples spermatozoa, spermatids, and spermatocytes. Each treated male should be assigned a unique identification number written on the vial which identifies the females inseminated by that male. In this way the progeny of each treated male can be kept separate and the data recorded in such a way that the origin of each tested chromosome is known.

The F_1 males should be collected and mated individually in single vials to virgin "Oster, *bw st p^p^"* females. An approximately equal number of F_1 males per treated male should be used for this cross to avoid biasing the data.

Scoring of the F_2 generation should be performed by looking for the presence or absence of flies (regardless of sex) with brown or orange eyes. The absence of these classes among at least 20 F_2 progeny indicates that the father (F_1 male) is heterozygous for a reciprocal translocation. If there are less than 20 flies per vial, the test cross should be repeated by mating three red-eyed F_2 males individually to virgin "Oster, *bw st p^p^"* females and scoring the progeny for flies with brown or orange eyes.

Data Analysis

The number of translocations and the number of gametes tested should be recorded for each brood. The frequency of translocation should be calcu-

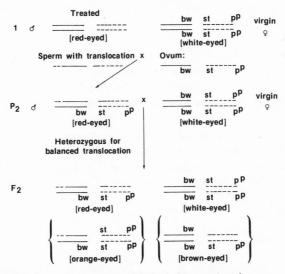

FIGURE 9.5. Mating scheme for the *Drosophila* heritable translocation assay. [ ], phenotypes of the flies; { }, unbalanced, inviable genotypes if translocation has occurred; ——, second chromosome; – – – –, third chromosome. Genetic markers: *bw*, recessive, light-brown eyes; *st*, recessive, bright red eyes (scarlet); *pP*, recessive, pink allele of peach gene.

lated for each brood as well as for each dose level. The data are then subjected to the Kastenbaum–Bowman statistical test to determine the significance of the results.

Reference

Lewis, E. B., and Bacher, F.: Method of feeding ethylmethane sulfonate (EMS) to *Drosophila* males. *Dros. Inf. Service* **48**:193, 1968.

PROTOCOL 21

SEX-LINKED RECESSIVE LETHAL TEST IN *D. MELANOGASTER*

The SLRL test in *Drosophila* provides a short-term *in vivo* mutagen screening system in a eukaryotic organism. The test measures the frequency of lethal mutations in approximately one-fifth the total genome of the fly. Accumulated evidence has indicated this test is the most sensitive among the different test systems available in *D. melanogaster* (Würgler *et al.*, 1975).

Materials

The stocks utilized for this test should be chosen for unique chromosome structure and visible markers. They should be maintained as permanent cultures and visually examined on a routine basis for any indication of genetic breakdown.

EMS at 0.015 M in 1% sucrose solution can be administered as the positive control as a single, 24-hr exposure on the final day of the dosing schedule. The males dosed should be the same age and stock as those used for the test article. The solvent or vehicle used for the test article should be used as the negative control and should be administered concurrently with the test article.

Drosophila cultures should be maintained at ~25°C in disposable vials. The culture medium may be Carolina Biological Instant *Drosophila* medium (Formula 4-24, without dyes) or the equivalent. Flies should be immobilized with filtered CO_2 for handling.

Experimental Design

Solubility Testing

Each test article should be tested for solubility; the primary choice for a solvent should be distilled water. If a test article is insoluble in water, it should be dissolved in DMSO, ethanol, acetic acid, or acetone and then diluted. The final concentration of solvent should be 2% or less.

Palatability and Toxicity Testing

A test for palatability and toxicity should be performed by preparing a test article in distilled water containing sucrose and any necessary solvents. The solution should then be administered by the following method modified from Lewis and Bacher (1968). A piece of chemically inert glass filter paper lining a shell vial should be saturated with 1.5 ml of the test solution. Fifty adult males should then be placed in each feeding vial and observed for feeding behavior and toxic effects. To ensure uptake it may be necessary to use concentrations of sucrose varying between 1 and 5% and to vary the feeding time between 1 and 3 days. It will be useful at this point to include food coloring in the feeding solutions of certain vials, as the coloring is readily distinguishable in the gut and feces of *Drosophila* and it can easily be determined that the flies have ingested the test article. Vials containing the coloring agent should not be used for assessment of toxicity. Should it be necessary to extend the dosing period beyond 24 hr, the test article should be reprepared and the flies transferred to a new feeding vial.

Fertility Testing

Once the solubility and toxicity have been established, a small pilot study should be conducted to determine the fertility of the treated males. Selected concentrations of the test article should be prepared, and the dosed males put through the same mating sequence as that to be used in the actual test to ensure that sufficient numbers of progeny will be available for analysis.

Test Size

The tests should be conducted at two dose levels in conjunction with negative and positive controls. At least 200 treated males should be used for each dose level of the test article and negative control. A suggested number of chromosomes to be tested and a brooding pattern are shown in the following table:

Chromosomes Tested

| | Days posttreatment | | | |
| | Brood I | Brood II | Brood III | |
Tests	1, 2	3, 4, 5	6, 7, 8	Total
Low concentration	2000	2000	2000	6000
High concentration	2000	2000	2000	6000
Negative control	2000	2000	2000	6000
Positive control	100	100	100	300

Procedure

Collection of Males

To provide males of approximately uniform age for treatment, males from culture bottles in optimum condition which are produced from timed (1-day) egg samples should be collected as virgins and held for a 2-day maturation period. All males should then be between 2 and 3 days of age and will have their full complement of sperm for treatment. Males should be randomized by the following unsystemized method. A sufficient number of healthy males for the test should be transferred to a single empty bottle, mixed, parcelled into groups of 50, and held in empty vials for several hours before exposure. This period of removal from a food supply should ensure immediate uptake of the feeding solution.

Dosing

The males should be dosed according to the methods described under Palatability and Toxicity Testing, using the concentration and duration of exposure determined in the preliminary studies. The high dose is normally set at one-half the LD_{50} and the low dose at one-quarter the LD_{50} unless contra-indicated by sterility of the dosed males. The maximum concentration administered, if the test article is nontoxic, should generally be 5%.

Mating and Scoring

The treated males should be mated individually to sequential groups of three virgin *Basc* (an acronym for Bar, apricot, scute) females. The brooding scheme should consist of a 2-3-3-day sequence which samples spermatozoa, spermatids, and spermatocytes. Each treated male should be assigned a unique identification number written onto the vial which identifies the females inseminated by that male throughout the brooding sequence. In this way the progeny of each male can be kept separate and the data recorded in such a way that the origin of each tested chromosome is known. This method eliminates the possibility of false positives resulting from clusters of identical lethal mutations originating in one treated male.

The F_1 progeny of each culture should be inspected to make certain the proper cross was made. The desired number of F_1 females should then be pair-mated to their brothers. An approximately equal number of F_1 females per treated male should be tested to avoid biasing the data.

In the F_2 generation, each culture vial (representing one treated X chromosome) should be examined for the presence of males with yellow bodies. If this class of males is present the culture should be considered nonlethal and discarded. If this class is absent the vial should be marked as a potential lethal and set aside for further examination. The following criteria should be applied to cultures suspected of being lethal:

1. If 20 or more progeny are present and there are no yellow-bodied males, the culture should be considered to carry a lethal mutation on the treated chromosome and further testing is not required. The chance of missing a male carrying the treated X chromosome in a population of this size is $(\frac{1}{2})^5$ or <0.05.

2. If there are fewer than 20 progeny or if there is one yellow-bodied male, the culture should be retested by mating three of the females heterozygous for the treated and *Basc* chromosomes to *Basc* males. The progeny of these crosses should be scored for the presence of *y* males.

The *Basc* (*Muller-5*) mating scheme employed in this test is shown below.

* This class of male is absent if a lethal mutation occurred on its X chromosome.
----, Treated X chromosome.

The treated X chromosome is marked with *y* (yellow body) for easy identification in the F_2 progeny since the other three genotypes have a normal black body. The X-chromosome balancer, *Basc*, carries a dominant *B* (Bar-eyed) gene which gives narrow eyes in a homozygous or hemizygous condition and kidney-shaped eyes in a heterozygous condition. The gene w^a (white apricot) is recessive and gives an apricot-eyed phenotype in the homozygous condition. The visible expression of the recessive gene *sc* (scute) is a variable reduction in the length or number of the thoraxal bristles.

Data Analysis

When the data are compiled, the total number of X chromosomes tested should equal the sum of the lethal and nonlethal cultures. The frequency of X-linked recessive lethals should be calculated as follows:

$$\frac{\text{number of lethals}}{\text{number of lethals} + \text{number of nonlethals}} \times 100 = \% \text{ lethal}$$

The Kastenbaum–Bowman test (Würgler *et al.*, 1975) should be used to determine the significance of the results. A significant increase of twice the spontaneous frequency in conjunction with a dose-response relation is considered a positive response.

In evaluating the results, particular attention should be given to multiple lethals occurring in progeny from a single treated male. In the positive control, where the mutation frequency is typically 20–25% SLRL, several lethals occurring in a single brood from one parent male can be the result of individual mutational events and should be counted as such in compiling the data. In the negative control, however, or in the test groups when the mutation frequency is very low, such occurrences will generally be interpreted as being in existence prior to treatment. The probability of a high number of independent events occurring by chance in one male should be calculated using a chi-square test. If it is determined that the unusual number of events, called a "cluster," cannot be attributed to handling or treatment, then the cluster should be determined to have arisen from a spontaneous mutation in a gonial cell which then replicated, and the cluster counted as a single event to avoid biasing the interpretation of results.

References

Lewis, E. B., and Bacher, F.: Method of feeding ethylmethane sulfonate (EMS) to *Drosophila* males. *Dros. Inf. Service* **48**:193, 1968.

Würgler, F. E., Graf, V., and Berchtold, W.: Statistical problems connected with the sex-linked recessive lethal test in *Drosophila melanogaster*. I. The use of the Kastenbaum/Bowman test. *Arch. Genet.* **48**:158–178, 1975.

Appendices

APPENDIX A: PREPARATION OF S9 LIVER HOMOGENATES

Induction of Microsomal Enzymes

Rats are typically used as the source of microsomal enzyme preparations for routine testing. Other species can be and have been successfully used, including humans. Small rodents are generally pretreated with enzyme inducers to elevate the level of microsomal activity. Agents such as polychlorinated biphenyls (Aroclor 1254), phenobarbital, or methylcholanthrene (MCA) give good induction for several sets of enzymes in the liver.[4]

The following steps should be followed when inducing animals for S9 preparation.

1. Animals should be quarantined at least 7 days prior to treatment and examined prior to use for evidence of disease.
2. Induction
   a. Phenobarbital (crystalline) should be suspended in corn oil (80 mg/kg body weight for rats, 160 mg/kg body weight for mice).[3] The phenobarbital should be administered I.P. once daily for 5 days, and the rodent sacrificed 24 hr after last injection. The animals should not be starved.[2]
   b. MCA should be suspended in corn oil (20 mg/kg body weight for rats, 40 mg/kg body weight for mice).[3] The chemical should be administered by I.P. injection once daily for 3 days. The animals should be sacrificed 24 hr after the last injection. They should not be starved.[2]
   c. Aroclor 1254 should be suspended in corn oil (500 mg/kg body

weight for rats, 1000 mg/kg body weight for mice).[3] The Aroclor 1254 should be administered I.P. once, and the animals sacrificed after five 24-hr periods. The animals should be starved 12 hr prior to sacrifice.[1]

3. Recordkeeping

All information related to animal husbandry, dates of injection, and sacrifice should be recorded on a form similar to the example shown in Figure A.1. Information from homogenization procedures can also be entered on this form.

Preparation of the S9 Fraction

After the animals have been treated for the appropriate period with the inducer, they should be killed and their livers removed for processing. Tissues should be homogenized in 0.15 M KCl or other suitable vehicle. Tissues other than liver have been used as activation sources, but in general they do not produce qualitative differences in compound activation.[6] Quantitative differences have been reported between mammalian species.[5]

The steps involved in S9 preparation are as follows (all KCl, homogenizing equipment, beakers, and tissue are kept as cold as possible at all times; tissue is handled aseptically).

1. The animal should be stunned by a cranial blow, then decapitated and bled before removing tissue.

2. Desired tissue should be excised using flame-sterilized instruments.

3. Excised tissue should be placed in a preweighed zip-lock storage bag or beaker and placed on ice.

4. Tissues and bag should be weighed to get the net tissue weight.

5. Tissue then should be rinsed in cold KCl solution.

6. Three milliliters cold (4°) KCl per gram of tissue should be measured to be homogenized. All solutions should be kept on ice.

7. The tissue and KCl should be placed in a tissue grinder tube. Potter–Elvehjem or Polytron homogenizer systems have been used with equivalent results.

8. The tissue grinder tube should be kept packed in ice while grinding. A constant grinding time is recommended for each run.

9. The homogenate should be decanted, keeping it chilled, and the process repeated until all the tissue is homogenized. Any unused KCl from the measured volume should be mixed in at this time.

10. The homogenate should be poured into centrifuge tubes and spun at 9000g for 20 min at 0°C.

11. The supernatant should be carefully removed and aliquoted into prelabeled, chilled vials.

12. A check for sterility should be made on complete and miminal agar plates.

Project#	Date Received
Purchase Request#	Date Induced
Purchase Order#	
Supplier	
Species/Sex	Date Food Removed
Mean Weight of Animals	Date Sacrificed
Lot#	
Number Induced	Number Sacrificed
Explain Any Loss of Animals	

Induction System	Supplier	Dose Level	Solvent
Aroclor 1234			
Phenobarbital			
3-Methyl-Cholanthrene			

Dose Level Calculations:

Date Homogenized	
Weight of Tissue Removed	
Homogenizing Solution Employed	Ratio
Volume of Homogenizing Solution Used	
Temperature of Centrifuge	
RPM Setting	Duration
Number of Vials/Volume in Each	
Final Volume Obtained	
Technician	Date

FIGURE A.1. Enzyme induction information data sheet (S-9).

13. The samples should be frozen at $-80°C$.

Samples may be stored at $-80°C$ for several months.

Preparation of the S9 Mix

The actual working S9 mixture should be prepared fresh and maintained at 4°C during use. There is no significant decay of activity over 4–5 hr at 4°C.

The following stock solutions should be prepared:

0.2 M NADP (filter sterilize)
0.2 M Glucose-6-PO₄ (filter sterilize)
0.4 M MgCl₂ (heat sterilize)
1.65 M KCl (heat sterilize)
0.2 M Sodium phosphate buffer, pH 7.4 (sterilize)

Three solutions (A, B, and C) can be made and stored in the following manner:

A. 6-ml aliquots of 0.2 M NADP can be dispensed into sterile vials and kept frozen at $-80°$ until used.

B. 7.5-ml aliquots of 0.2 M glucose-6-PO₄ can be dispensed into sterile vials and kept frozen at $-80°$ until used.

C. The following can be mixed and stored at 4°C.

0.4 M Cl₂	20 ml
1.65 M KCl	20 ml
0.2 M phosphate buffer	500 ml
Sterile distilled H₂O	315 ml

TABLE A.1
Complete Activation System Composition[a]

Components	Mol. wt.	Concentration/ml S9 mix
NADP	801	4 μmol
Glucose-6-PO₄	283	5 μmol
MgCl₂	203.3	8 μmol
KCl	74.6	33 μmol
Sodium phosphate buffer, pH 7.4		100 μmol
Organ homogenate (S9 fraction)		100 μl

[a] This mixture should be discarded after use and not refrozen. Several activity parameters of the S9 may be measured (e.g., specific enzyme activity and milligrams protein per milliliter, sterility, vibration against reference mutagens).

Ten milliliters of S9 mix can be prepared by adding:

$$\underline{\begin{array}{ll} 0.2 & \text{ml of solution A} \\ 0.25 & \text{ml of solution B} \\ 8.55 & \text{ml of solution C} \\ 1.00 & \text{ml of S9 fraction} \end{array}}$$

10.00 ml

The composition per milliliter of the mix is identified in Table A.1.

REFERENCES

1. Ames, B. N., McCann, J., and Yamasaki, E.: Methods for detecting carcinogens and mutagens with the *Salmonella*/mammalian-microsome mutagenicity test. *Mutat. Res.* **31**:347–364, 1975.
2. Burke, M. D., and Mayer, R. T.: Ethoxyresorufin: Direct fluorimetric assay of microsomal *O*-dealkylation which is preferentially inducible by 3-methylcholanthrene. *Drug Metab. Dispos.* **2**(6):583–588, 1974.
3. Freireich, E. J., *et al.*: Quantitative comparison of toxicity of anti-cancer agents in mouse, rat, dog, monkey and man. *Cancer Chemother. Rep.* **50**(4):219–244, 1966.
4. Litterst, C. L., Farber, T. M., Baker, A. M., and van Loon, E. J.: Effect of polychlorinated biphenyls on hepatic microsomal enzymes in the rat. *Toxicol. Appl. Pharmacol.* **23**:112–122, 1972.
5. Oesch, F., Raphael, D., Schwind, H., and Glatt, H. R.: Species differences in activating and inactivating enzymes related to the control of mutagenic metabolites. *Arch. Toxicol.* **39**:97–108, 1977.
6. Weekes, U.: Metabolism of Dimethylnitrosamine to mutagenic intermediates by kidney microsomal enzymes and correlation with reported host susceptibility to kidney tumors, *J. Natl. Cancer Inst.* **55**:1199–1201, 1975.

APPENDIX B: DOSE SELECTION FOR *IN VIVO* GENETIC ASSAYS

Selection of dose levels for *in vivo* studies is often one of the most critical factors in the entire operation. Dose levels which are suboptimal for the type of test being conducted could lead to erroneous conclusions when data are evaluated.

In general, genetic assays incorporate either acute or subchronic chemical administration. Exposure is commonly limited to a single or five daily administrations. Exceptions will be the mouse heritable translocation assay or the extended dosing modification for the dominant lethal assay in which the chemical is administered continuously over the spermatogenic cycle of the male.

The common practice for dose determination is to perform a dose range experiment prior to the actual study. This usually consists of setting up five to seven groups of animals with four to eight animals per group. A series of doses is given to the groups on an acute basis using the anticipated route of exposure. If nothing is known about the toxicity of the test material, the maximum dose in the series might be 10 g/kg body weight with lower doses in the series falling at natural logarithmic spaces or steps (e.g., 10 g/kg, 3 g/kg, 1 g/kg, 0.3 g/kg, etc.).

Using either lethality or some other sentinel parameter, an LD_{50} or ED_{50} (effective dose 50%) level can be calculated by one of several methods.[2,3,4]

If the duration of exposure in the study is short and the animals are going to be evaluated shortly after exposure (6–48 hr), then a dose near the LD_{50} or ED_{50} level should be selected (e.g., LD_{50}).

This maximum is considered the maximum tolerated dose (MTD) over the course of the study* and subjects the animals to levels of the chemical in excess of what might be expected in the environment barring accidental release. It may be interpreted as the worst possible case. Typically, one or more lower doses are included in the study design, which permits evaluation at exposures near the anticipated use or environmental levels. If the data of a study using this dose selection scheme are negative, confidence can be had in extrapolating the results to larger populations.

If the data from the study show an effect, the shape of the response curve over several dose levels can be of importance in interpreting the likely effects near the anticipated use or environmental level. It is almost impossible to develop reliable interpretations from a study if the anticipated use or environmental level is used as the maximum dose, since all *in vivo* assays are characterized by limited sampling. Agents with weak to moderate activity could not be detected under such conditions.

When the dosing period is extended to 5 days or longer, it is very likely that the MTD will be smaller. This is not true for some relatively inocuous chemicals, but does apply to the majority of chemicals. The reasons are varied but include:

1. Incomplete clearance of the chemical between administrations, thus increasing the total daily dose and leading to unanticipated toxicity.
2. Chemical interactions may lead to enzyme induction and possible alterations in metabolism with concomitant changes in toxicity.
3. Some chemicals produce delayed toxicity that is not observed in studies covering one or a few days, but is evident in more prolonged exposures.
4. Some chemicals may modify the animals' immume systems, leading to hypersensitivity or immune suppression. Either result can significantly alter the toxic effects of the chemical during extended dosing.

* The MTD is expected to be minimally toxic to the test animals over the course of the study.

Therefore, before initiating studies such as the mouse heritable translocation assay which may require 7–10 weeks' exposure, it will be necessary to conduct a long-term pilot study at several doses or reduce the acute MTD by some factor. For example, it is not uncommon to use $\frac{1}{8}$–$\frac{1}{10}$ LD_{50} as the high dose in an extended exposure study design. Mid- and low-dose levels are usually selected as some fraction of the high dose (e.g., $\frac{1}{3}$ and $\frac{1}{10}$, respectively). This type of factoring is risky, since significant toxicity or lethality may result well into the study. A long-term pilot study, although costly, could prevent loss of the entire study and is the recommended method for dose selection.

Because so many of the *in vivo* assays for genetic end points involve mating sequences, it may be necessary to use an end point other than lethality to select the MTD. Animals with suppressed libido or nonviable sperm will not provide an accurate assessment of heritable genetic end points. Calculation of an ED_{50} for a parameter such as fertility or neurological effects might be appropriate for some chemicals. Regardless of the type of study, route of exposure, or duration of compound administration, it will be necessary to make regular (preferably twice daily) inspection for toxic signs.[1] Recording toxic signs is often an aid in the evaluation of the results, since it gives another parameter to the toxicity of the compound. Unusual findings at animal necropsy should also be recorded in the raw data.

REFERENCES

1. Campbell, P. E. S., and Richter, W.: An observation method estimating toxicity and drug actions in mice applied to 68 reference drugs. *Acta Pharmacol. Toxicol.* **24**:345–63, 1967.
2. Finney, D. J.: Statistical Method in Biological Assay. Hafner Press, New York, 1969.
3. Litchfield, J. T., and Wilcoxson, F.: A simplified method of evaluating dose-effect experiments. *J. Pharmacol. Exp. Ther.* **96**:99–113, 1949.
4. Miller, L. C., and Tainter, M. L.: Estimation of the ED_{50} and its error by means of logarithmic-probit graph paper. *Proc. Loc. Exp. Bid. Med.* **57**:621–64, 1944.

APPENDIX C: SELECTED REFERENCES AND REVIEWS OF GENETICS AND GENETIC TOXICOLOGY*

Chemical Mutagenesis in Mammals and Man (F. Vogel and G. Röhrborn, eds.), Springer-Verlag, New York, 1970.

---

* Additional information may be available through the Environmental Mutagen Information Center (EMIC) at the Oak Ridge National Laboratories. This service maintains many manuscripts and publications on environmental mutagens (excluding radiation alone) that have been printed since 1969 and it is able to answer bibliographic and reference questions. The service is directed by Mr. John Wasson at Environmental Mutagen Information Center, Oak Ridge National Laboratory, Building 9224, Oak Ridge, TN 37830. Telephone: (615) 574-7871.

Chemical Mutagens (L. Fishbein, W. G. Flamm, and H. L. Falk), Academic Press, New York, 1970.

Chemical Mutagens: Principles and Methods for Their Detection, Vols. 1–4 (A. Hollaender, ed.); Vol. 5 (A. Hollaender and F. J. de Serres, eds.); and Vol. 6 (F. J. de Serres and A. Hollaender, eds.), Plenum Press, New York, 1971–1979.

Genetics (M. Strickberger), Macmillan, New York, 1968.

Handbook of Mutagenicity Test Procedures (B. J. Kilbey, M. Legator, W. Nichols, and C. Ramel, eds.), Elsevier, Amsterdam, 1977.

Human Genetics, E. Novitski. Macmillan, New York, 1977.

Molecular Biology of the Gene, J. Watson. W. A. Benjamin, New York, 1970.

Progress in Genetic Toxicology (D. Scott, B. A. Bridges, and F. H. Sobels, eds.), Elsevier/North-Holland, Amsterdam, 1977.

Strategies for Short-Term Testing for Mutagens/Carcinogens (B. E. Butterworth and L. Golberg, eds.), CRC Press, West Palm Beach, Fla., 1979.

Index